TECHNIQUES IN RHEOLOGICAL MEASUREMENT

TECHNIQUES IN RHEOLOGICAL MEASUREMENT

Edited by

A. A. COLLYER

Division of Applied Physics,
School of Science,
Sheffield Hallam University, UK

CHAPMAN & HALL

London · Glasgow · New York · Tokyo · Melbourne · Madras

Published by Chapman & Hall, 2-6 Boundary Row, London SE1 8HN

Chapman & Hall, 2-6 Boundary Row, London SE1 8HN, UK

Blackie Academic & Professional, Wester Cleddens Road, Bishopbriggs, Glasgow G64 2NZ, UK

Chapman & Hall, 29 West 35th Street, New York NY10001, USA

Chapman & Hall Japan, Thomson Publishing Japan, Hirakawacho Nemoto Building, 6F, 1-7-11 Hirakawa-cho, Chiyoda-ku, Tokyo 102, Japan

Chapman & Hall Australia, Thomas Nelson Australia, 102 Dodds Street, South Melbourne, Victoria 3205, Australia

Chapman & Hall India, R. Seshadri, 32 Second Main Road, CIT East, Madras 600 035, India

First edition 1993

© 1993 Chapman & Hall

Typeset in Times 10/12pt by J.W. Arrowsmith Ltd
Printed in Great Britain by the University Press, Cambridge

ISBN 0 412 53490 8

A catalogue record for this book is available from the British Library
Library of Congress Cataloging-in-Publication Data available

Preface

In an earlier book, *Rheological Measurement* (A. A. Collyer & D. W. Clegg, Elsevier Applied Science, 1988), the basic rheological methods of measurement presently used were discussed in the light of the basic underlying principles and current theories. The same approach is adopted in this companion book, which is concerned with some newer or more sophisticated techniques that have resulted from a fresh understanding of the subject, or as a result of improvement in computer control, data acquisition and computational power, or more simply from an industrial need, particularly with regard to process control.

The first two chapters deal with the extensional flow properties of fluids and their measurement. This inclusion is in response to a greater awareness in industry of the importance of these flows. Chapter 3 introduces and develops the subject of surface rheology and the measurement of its properties, again a subject of increasing significance. The methods of measurement of the dynamic mechanical properties of fluids and the calculation of the resulting rheological parameters are discussed in Chapters 4–7 inclusive. The subject areas covered are: large-amplitude oscillatory shear, a model for viscoelastic fluids and solids, a new method of measuring dynamic mechanical properties, particularly for curing systems, and the use of complex waveforms in dynamic mechanical analysis. Rheological measurements on small samples, typical of those obtained from biological systems or from new chemical syntheses, are described in Chapter 8. The last two chapters are of relevance to the measurement of rheological parameters during processing. The topics discussed are speed- or stress-controlled rheometry and rheometry for process control.

It is hoped that this book will be a suitable introduction to those rapidly changing areas in rheological measurement, and make readers more aware of the greater variety of techniques that can be used to assist in the understanding of the way in which their fluids are behaving during processing or in quality control, be they polymer melts, biological materials, slurries, or food materials. Where mathematics is used, the reader need have no further knowledge than 'A' level or a pre-university level.

This work is of importance to all establishments in which rheological work is carried out. Material scientists, engineers, or technologists in industry, research laboratories, and academia should find this book invaluable in updating their information and understanding of the wide-ranging area that is rheology.

A. A. COLLYER

Contents

List of Contributors

M. BAUMGÄRTEL
Department of Chemical Engineering, University of Massachusetts, 159 Goessmann Laboratory, Amherst, Massachusetts 01003, USA

D. BINDING
Department of Mathematics, UCW, Penglais, Aberystwyth, Dyfed, SY23 3BZ UK

T. O. BROADHEAD
Department of Chemical Engineering, McGill University, 3480 University St., Montreal, Quebec, Canada H3A 2A7

J. M. DEALY
Department of Chemical Engineering, McGill University, 3480 University St., Montreal, Quebec, Canada H3A 2A7

A. J. GIACOMIN
Department of Mechanical Engineering, Texas A&M University, College Station, Texas 77843-3123, USA

W. GLEIẞLE
Institut für Mechanische Verfahrenstechnik und Mechanik, Universität Karlsruhe (TH), Kaiserstrasse 12, D-7500 Karlsruhe, Germany

D. F. JAMES
Department of Mechanical Engineering, University of Toronto, Toronto, Ontario, Canada, M5S 1AL

M. E. MACKAY
Department of Chemical Engineering, University of Queensland, St. Lucia, Brisbane, Queensland, Australia 4072

B. I. NELSON
Department of Chemical Engineering, McGill University, 3480 University St., Montreal, Quebec, Canada H3A 2A7

R. A. PETHRICK
Department of Pure and Applied Chemistry, University of Strathclyde, Thomas Graham Building, 295 Cathedral St., Glasgow, UK, G1 1XL

P. R. SOSKEY
ENICHEM Americas Inc., 2000 Princeton Park, Corporate Center, Monmouth Junction, New Jersey 08852, USA

K. WALTERS
Department of Mathematics, UCW, Penglais, Aberystwyth, Dyfed, SY23 3BZ UK

B. WARBURTON
School of Pharmacy, University of London, 29/39 Brunswick Square, London, UK, WC1N 1AX

H. H. WINTER
Department of Chemical Engineering, University of Massachusetts, 159 Goessmann Laboratory, Amherst, Massachusetts 01003, USA

Chapter 1

Contraction Flows and New Theories for Estimating Extensional Viscosity

D. M. BINDING

Department of Mathematics, University of Wales, Aberystwyth, UK

1.1. INTRODUCTION

A knowledge of the extensional viscosity of a fluid, or at least a quantity that reflects its extensional properties, is crucial to the overall understanding of how a fluid will respond in different flow situations. The extensional viscosities, for example, can be several orders of magnitude higher than the corresponding shear viscosities, and this can have a dramatic influence on the flow field in a complex process.

The measurement of extensional viscosity, however, is not a straight-forward task, particularly for mobile fluid systems. This fact arises principally because of the difficulty encountered in generating a well-defined extensional flow field in the fluid. Experimentally, it is simply not possible to apply to a fluid the relevant boundary conditions necessary to sustain such a flow. Chapter 2 of this book refers to many of the problems involved. Invariably one has to resort to studying flows that are known to contain a substantial component of extension in order to extract from them the required information. Contraction flows are an example that satisfies that need.

The contraction flow problem is a fundamentally important one in the field of rheology. With the exception of shear flow it has probably received more attention from researchers than any other flow and is the subject of regular reviews (recent ones include those by White *et al.*[1] and Boger[2]).

Such devotion to one particular flow is easy to understand. Analytically the problem is insoluble even for the simplest of materials such as Newtonian fluids. As a consequence, therefore, many techniques such as simple approximations to the velocity field, boundary-layer analysis, variational methods, asymptotic expansions, etc., have been used to provide useful information about various aspects of the flow.

Experimentally the problem is, at least in principle, quite straight-forward because of the geometric simplicity involved, and measurements of quantities such as excess entry pressure are now fairly routine. Determination of velocity and stress fields is more difficult, however, and reliable data are not plentiful in the literature. On the other hand, flow visualisation studies have unfolded a situation that is as complex as is likely to be encountered in any flow problem. Observations of vortex enhancement and, more recently, of the generation of 'lip' vortices as well as several other unusual flow features have provided further impetus to studies of entry flows.

This diversity of intriguing flow behaviours has provided the numerical analysts with a geometrically simple problem of considerable kinematic complexity, on which to test a multitude of numerical schemes. For good measure the problem provides singular points in the flow field that require particular attention. Unfortunately, the difficulties are such that they have detracted somewhat from one of the important aims of numerical simulation studies, that of differentiating between the many viable constitutive relations used to model the fluids' responses. Very recently, however, significant advances have been made.

Tackling the entry flow problem is also extremely important from a practical point of view, since it closely resembles many flow processes that are encountered in the manufacturing industries. Of equal importance in this respect are exit flows, although these have not received the same degree of attention that entry flows have enjoyed. This is perhaps understandable, since they do not generally exhibit the same array of fascinating features as their entry counterparts and, therefore, do not offer as great a challenge or interest to the analysts. However, such flows have not been neglected. Indeed, exit flows have provided their fair share of controversy as the review by Boger and Denn[3] testifies.

Given the experimental difficulties involved in measuring the extensional viscosity of mobile fluids (see, for example, Chapter 2 of this book), contraction and converging flows are increasingly being used for that purpose. Thus it is essential to understand how a fluid's rheometric properties interact and manifest themselves in such flows. The emphasis in this chapter will be, therefore, on providing a unified, albeit personal, picture of entry flows, with particular attention given to examining the influence of shear viscosity, extensional viscosity and elasticity on the observed behaviour. For the sake of clarity and general convenience, attention will be restricted to axisymmetric flows.

1.2. THE CONTRACTION FLOW PROBLEM

The traditional contraction flow problem is depicted schematically in Fig. 1.1. Fluid flows through a long circular cylinder of diameter D, the axis of which is taken to be the z-direction of a cylindrical coordinate system (r, θ, z). At $z = 0$ the cylinder abruptly and symmetrically contracts into a similar cylinder of diameter d. The ratio of the two diameters, $a = D/d$, is referred to as the contraction ratio. Both cylinders are assumed to be sufficiently long for fully developed conditions to exist at both ends of the geometry. The fully developed regions are of little interest as they are fully understood, and so attention is focused on the influence of the contraction on the flow immediately upstream and downstream of the contraction plane. This then is commonly referred to as 'entry flow'.

Two lengths have been identified as being important parameters in entry flows: the vortex length L_v is the distance upstream of the contraction plane at which the observed secondary flow detaches from the cylinder wall, and L_e is the distance downstream of the contraction plane

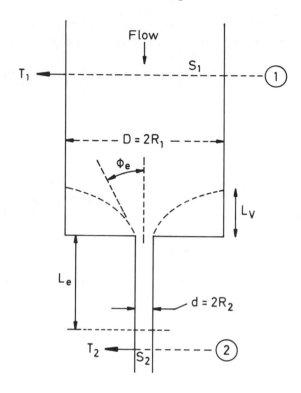

Fig. 1.1. Schematic diagram of the contraction flow problem.

required for the centre-line velocity to reach 90%, say, of its fully developed value. Additionally, the entrance angle ϕ_e, the angle that the vortex makes relative to the flow direction at the lip of the contraction, is also sometimes reported.

At this point some observations are in order. First, a third length, namely the distance upstream of the contraction plane at which the flow starts to depart from fully developed conditions (an exit length?), would also seem to be an important parameter that has received scant attention in the literature. A further point worth noting is that both vortex and entry lengths are characteristics of the kinematics of the flow and do not necessarily reflect the behaviour of the stress field. Both of the above points are clearly relevant to the design of, and subsequent interpretation of data from, instruments such as capillary rheometers. In this respect vortex and entry lengths may not be adequate measures.

The technique for extracting the shear viscosity as a function of shear rate from measured contraction flow pressure drops is well established and not controversial. Provided that the so-called Bagley correction is applied correctly, the interpretation of data is straightforward (see, for example, Ref. 4). Nonetheless, it is worth mentioning that the only strictly valid Bagley procedure requires the pressure drop to be measured for two capillaries of different lengths *but having the same diameter* (assuming the upstream cylinder diameter to be constant) and under such conditions that for both capillaries, both upstream and downstream, fully developed regions are present. Many commercial capillary rheometers have flows that are driven by means of a constant-speed piston and it is not difficult to imagine, particularly at high speeds and near the end of the piston stroke, that the available upstream distance may be insufficient to guarantee fully developed conditions.

What has concerned researchers recently has been the accurate determination of the entry pressure drop (total measured pressure drop less that expected on the basis of fully developed Poiseuille flow) and the dependence of this quantity upon parameters such as the extensional viscosity, as well as the study of other associated phenomena that occur at the capillary entrance. The non-dimensionalised entry pressure drop will be referred to as the Couette correction, but there is some confusion in the literature as to the precise definition of this parameter. It is hoped that the situation will be clarified during the course of this chapter.

1.3. EXPERIMENT

1.3.1. Experimental Technique

Experimentally, contraction flows are relatively easy to generate and study. A typical experimental set-up is shown schematically in Fig. 1.2. A fluid sample is pumped at a constant flow rate from a reservoir, through the contraction geometry and then back to the reservoir. Depending upon the type of pump used, a means of smoothing the flow may be necessary. The geometry itself is often made of a transparent material to facilitate flow visualisation studies.

Flow rates are readily determined. Simple catch and measure techniques are, perhaps, the most common, although centrifugal mass-flowmeters can be used for low-viscosity fluids. Alternatively, positive displacement pump systems allow the flow rates to be calculated directly.

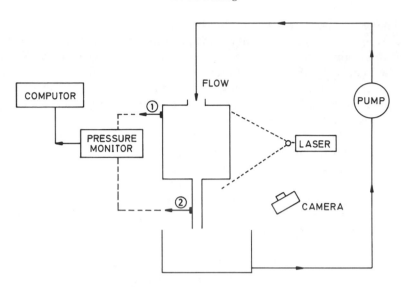

Fig. 1.2. Schematic diagram of a typical experimental set-up for studying contraction flows.

Flow visualisation can often produce information that is more enlightening than measured data for particular quantities. Certainly it is fair to say that such studies have provided much of the impetus that now exists in the area of numerical simulation of complex flows. Various visualisation techniques have been used, the most common for polymer solutions being that of speckle photography. For this, the fluid is 'seeded' with a small quantity of very small diameter particles such as polystyrene dust. A monochromatic source of light, usually from a low-power laser, is passed through a circular cylindrical lens to produce a single plane of light that passes through the geometry. The reflected light from the seeding particles can then be photographed. Long-time exposures of steady flows produce streamline patterns for the particular plane of the flow that is illuminated. Shorter time exposures can be used to obtain measurements of the velocity field. Laser-Doppler anemometry is also sometimes used for measuring the velocity field.

Birefringence techniques have been used to measure stress fields in the flow but it is more usual to measure a single stress on the wall of the upstream cylinder. This is done by means of a pressure transducer[1] mounted as flush to the upstream cylinder wall as is practically possible, at a location far enough upstream of the contraction plane for the flow there to be considered fully developed. Figure 1.2 indicates an ideal situaation in which a second transducer②is placed in the downstream cylinder. Although highly desirable, such a measurement is difficult to obtain since the downstream geometry radius is usually too small to accommodate a transducer without disturbing the flow. In practice, assumptions have to be made about the relevant exit stress conditions.

1.3.2. The Flow Field
Figure 1.3 contains schematic representations of streamline patterns obtained by various workers under a variety of conditions. The observed and predicted pattern for Newtonian fluids (Fig. 1.3(a)) contains a small recirculating vortex ($L_v = 0.17D$) that is essentially independent of Reynolds' number, although it gradually diminishes in size as fluid inertia begins to dominate.[5,6] Also significant is the fact that the entry plane velocity profile develops concavities at high Reynolds' numbers. Further

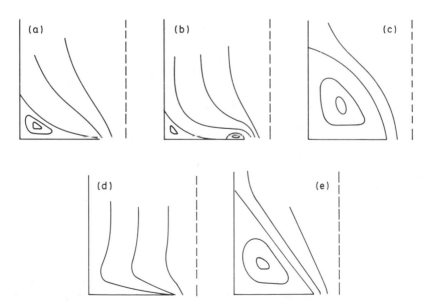

Fig. 1.3. Schematic representations of experimentally observed entry flow patterns.

details of the Newtonian as well as the inelastic contraction flow problems, are contained in Boger's work.[2]

At sufficiently low flow rates, all fluids exhibit the Newtonian streamline pattern. However, at higher flow rates a variety of different patterns may be observed.[7–25] The most common one is the large recirculating vortex shown in Fig. 1.3(c) which generally increases in size with increasing flow rate, a phenomenon known as vortex enhancement (see, for example, Ref. 8). The fluid flows through the central 'funnel' which becomes more elongated as the vortex enlarges.

Of particular interest recently has been the transition from the Newtonian streamline pattern to that depicted in Fig. 1.3(c). The transition is often (but apparently not always) accompanied by the appearance of a lip vortex (cf. Refs 5 and 8–10) that coexists, for a limited range of flow rates, with the Newtonian salient corner vortex, as shown in Fig. 1.3(b). Several of the studies have looked at the effect of parameters such as contraction ratio, type of contraction, smoothness of entry and stability of flow. It is now generally accepted that vortex enhancement 'is very complex and is strongly affected by such factors as the rheological properties of the test liquids, the contraction ratio, the convergence angle, the sharpness of the reentrant corner, and the flow rate'.[8]

Several other intriguing patterns can be obtained. The 'bulb' shape pattern shown in Fig. 1.3(d) has been observed for constant-viscosity elastic fluids flowing through planar contractions.[11] Straight-edged vortices (Fig. 1.3(e)) are typical for fibre-filled fluids and also for solutions of polymers with semi-rigid rod-like molecules.[12,13] These results probably have a significance that, as yet, is not fully appreciated.

Each of the patterns in Fig. 1.3 may be thought of as representing relief mechanisms for the stress, in the sense that alternative velocity fields would have associated with them higher stresses and stress gradients and as a consequence would require a larger amount of work done to sustain the flow.

Visualisation studies with flowing polymer melts is much more difficult. Those that have been made demonstrate the existence of large recirculating vortices for some polymers, whereas others do not exhibit such vortices (cf. Refs 16–21) but clearly the scope for more detailed visualisation is limited.

There have been relatively few studies in which direct measurements of the velocity field have been made. Metzner *et al.*[26] and Binnington *et al.*[27] have produced limited data using laser-speckle photographic techniques. Laser–Doppler velocimetry has been used in several studies. Xu *et al.*[28]

showed the development of off-centre maxima in the velocity profile for a solution of silicone rubber in silicone oil, flowing at only moderate Reynolds' number through a rectangular slit. Hasegawa and Iwaida[29] studied the flow of dilute aqueous solutions of polyethyleneoxide and polyacrylamide through small apertures. Lawler *et al.*[30] (see also Ref. 31) studied several fluids in an axisymmetric contraction flow. For a PIB solution, two time-independent regimes were identified, separated by a time-dependent region corresponding to the transition from the low flow-rate pattern of Fig. 1.3(a) to the high flow-rate situation of Fig. 1.3(c). Wunderlich *et al.*[32] found off-centre maxima in the flow of a dilute solution of polyacrylamide through a planar contraction. Interestingly, divergent flow was also detected upstream of the contraction plane. Finally, Shirakashi and Watanabe[33] studied the effect of the upstream channel height on both pressure and velocity fields.

1.3.3. The Stress Field
Flow birefringence studies have been used to determine stress fields in contraction flows. Boles *et al.*[34] found very high shear-stress gradients in the converging region for a concentrated polyisobutylene solution flowing through a planar contraction, the concentration of stresses being nearer the centre-line than the geometry walls. Interestingly, the shear-stress levels were very similar to the levels of the normal stress difference. They also observed stress discontinuities both upstream and downstream of the contraction plane. Eisenbrand and Goddard[35] studied the flow of an aqueous solution of polyacrylamide through orifices. They concluded that although the birefringence correlated with the Weissenberg number, the measured pressure drops did not. Using speckle photography, they also measured local stretch rates which turned out to be considerably lower than had been anticipated.

Most studies report the total pressure drop required to produce a given flow rate through the geometry. In addition to many of the studies already mentioned, pressure-drop measurements have also been reported in numerous other papers. References 36–48 list some of those studies, the majority of which indicate an entry pressure drop substantially in excess of that expected on the basis of inelastic flow.

1.3.4. Numerical Simulation
The purpose here is to highlight some of the recent developments in the numerical simulation of contraction flows that shed light on the influence

of the rheometric properties on the observed macroscopic features as described above. More detailed reviews of the problem are available and the interested reader is referred to Refs 49–51.

The earlier work of Kim-E *et al.*[6] for axisymmetric contractions clearly demonstrated that increased shear-thinning in inelastic fluids has the effect of reducing vortex size, while at the same time increasing the entry pressure loss. Inertial effects also act to reduce vortex size.

Recent numerical simulation studies of the entry flow of elastic fluids appear to suggest that many of the earlier difficulties, resulting in lack of convergence at low values of the Weissenberg number, have been overcome. Marchal and Crochet[52] have obtained substantial vortex enhancement in a 4–1 contraction for an Oldroyd B model, in qualitative (but not quantitative) agreement with experimental data. Debbaut *et al.*[53] have extended this to other models such as the Phan Thien–Tanner and Giesekus–Leonov with similar success and, in addition, calculated that the Couette correction passes through a minimum, after which substantially increased values are obtained. They also attempted to relate changes in the macroscopic features with differences in the rheometric properties as predicted by the models. Apart from the obvious non-linear relationships, the rheometric parameters cannot be varied independently, making such a task very difficult. Dupont and Crochet[54] and Luo and Mitsoulis[55] used a KBKZ model to successfully predict the vortex growth of an LDPE melt, as well as the rapidly increasing Couette correction.

In a series of 'experimental' simulation studies, Crochet and colleagues[56,57] have employed, amongst others, a generalised Newtonian model with a viscosity function dependent upon both the second and third rate of strain invariants. Thus they arrive at the conclusion that elasticity (as manifested by the first normal stress difference in shear) results in a reduction of the Couette correction as well as a decrease in vortex size. High Trouton ratios, on the other hand, have the opposite effect, i.e. they lead to vortex enhancement and an increasing Couette correction. Debbaut[58] has recently obtained simulations of the generation of lip vortices and looked at the inertial and extensional effects on both corner and lip vortices. Debbaut[59] has also looked at the effect of stress overshoot and compared modified Oldroyd B and phan Thien–Tanner models.

Using a finite difference scheme, Chiba *et al.*[60] successfully simulated the axisymmetric contraction flow of dilute suspensions of rigid fibres in Newtonian fluids. Characteristic straight-sided vortices were obtained, the lengths of which were essentially independent of flow rate. Song and

Yoo,[61] using a finite difference scheme, studied the Upper Convected Maxwell fluid flowing in a 4–1 planar contraction. The results indicate a reducing vortex size with increasing Reynolds' number. The same model was used by Rosenberg and Keunings[62] in a finite element scheme, but here the emphasis was on problems of convergence and the existence or not of real limit points. Finally, the results of Choi *et al.*[63] for a Giesekus fluid in a planar contraction suggest that corner vortex growth decreases (increases) with small (large) elasticity number, in the case of a non-vanishing Reynolds' number.

1.3.5. Discussion

It is clear from the above review that basic rheometrical properties influence entry flow characteristics in a very complex manner. Certainly, the extensional viscosity can have a dramatic effect on the velocity field. This may have a bearing on the interpretation of data from experiments in which the velocity field is assumed *a priori* to adopt a specific form. Problems that could arise include the formation of secondary motions, the unexpected appearance of 'irregular' velocity profiles or even the onset of instabilities as the fluid attempts to avoid the build-up of large stresses.

The fact that numerical studies indicate an initial reduction in the value of the Couette correction, which has not been shown conclusively in experimental studies, is almost certainly due to the difficulties encountered in measuring the very small pressure differences that are involved at low values of the flow rate. Also, there is now little doubt that the reduction is not so much a characteristic of the flow; rather it is a consequence of the normal stresses existing at the exit plane and the way in which the Couette correction is defined.

For many years the idea that shear-flow experiments could explain all of the phenomena associated with contraction flows prevailed over the attempts by some researchers to introduce other explanations. This is reflected in the literature by the extensive use of Weissenberg or Deborah numbers for correlating experimental data. Both of these parameters incorporate a characteristic time that is usually determined from a shear-flow experiment.

Indeed, Gleissle[64] recently expounded the apparent mutual proportionality between the excess entry pressure and the first normal stress difference in shear. Such empirical relationships may well be valid for restricted classes of material, but numerical evidence suggests that the correlations are perhaps somewhat fortuitous. Some commercial instruments utilise empirical relationships of this type in order to extend the range of information obtainable from the apparatus and so it is clearly necessary to determine the limits for their applicability.

Even after extensional viscosity was suggested as being an important parameter in contraction flows, there appeared to be a reluctance in the literature to accept the fact that extensional viscosity was not simply a manifestation of elasticity. That this is the case is not difficult to show. Firstly, inelastic constitutive models can be generated in such a way that they not only satisfy all the accepted principles of continuum mechanics but also exhibit arbitrary extensional viscosity behaviour away from the limits of low strain rates. As was mentioned above, such models have been employed in numerical simulation studies, although they are obviously not recommended for general use.

Secondly, there exist perfectly acceptable physical mechanisms by which exceptionally large extensional viscosities in solutions can be generated. One such mechanism is that described by Batchelor[65] and Goddard[66] in which large aspect-ratio, inextensible fibres are suspended in an inelastic solvent. The excess stresses in an extensional flow arise from the fact that the fibres cannot flow affinely with the macroscopic flow and therefore generate high shear stresses in the sleeve of solvent surrounding each fibre. The resulting effective extensional viscosity is then determined essentially by the shear viscosity of the solvent and the geometric characteristics of the fibres. No parameter associated with elasticity is involved. On the other hand, the elasticity of the solution will clearly be dependent upon the mechanical properties of the fibres which can be varied independently of their geometrical properties. It is also interesting to note that in the case of solvents with constant or power-law shear viscosities, the predicted Trouton ratio for the suspension is independent of strain rate.

In the sections that follow it is hoped to show, at least qualitatively, how shear viscosity, extensional viscosity and elasticity *as independent fluid properties* all contribute in different ways to produce the various entry flow characteristics that have been described above. The analysis will be presented as fully as possible to illustrate the various assumptions and approximations that are made. It must be emphasised that, in order to achieve the stated aims, some of the mentioned approximations are physical rather than mathematical in nature. They are, however, clearly noted, so that they can be corroborated by further study or numerical simulation. Meanwhile, the usefulness of the underlying philosophy should be judged on the ability of the theory to reflect and explain the many complex features that have been observed in contraction flows.

1.4. ANALYSIS

1.4.1. General Considerations

One of the earliest attempts to describe mathematically the entrance effect was by Philipoff and Gaskins,[67] who separated the entrance correction into the sum of an inelastic 'geometrical end-correction' which they referred to as the 'Couette correction' and an elastic contribution which they associated directly with the recoverable strain as measured in a rotational steady-shear experiment.

There have been many attempts at relating entry pressure drop (or Couette correction) to fundamental rheometrical properties. Several analyses assume the flow to be essentially a shear-free sink flow[68-73] and relate the entry pressure drop directly to the extensional viscosity. Two principal criticisms can be levelled at those early analyses. The first is that the assumption of shear-free flow is unrealistic. Certainly, the experimental evidence does not support such an approximation. Also, it is felt that any approximation to the flow should also at least approximately satisfy the true boundary conditions relevant to the problem. The second, and possibly more severe, objection is that the analyses require, as input, information that should in fact form part of the solution to the problem. More specifically, the angle of convergence, or some other equivalent information, has to be determined experimentally before the analyses can be applied. A more recent analysis by Tremblay,[74] using an approximate sink flow as a starting point, also requires the input of a convergence angle or, alternatively, the acceptance of equivalent empirical data.

The only early analysis that escapes the above criticisms is that of Cogswell[75] who was the first to recognise the importance of extensional viscosity in contraction flows. Gibson and Williamson[43,44] and Kwon *et al.*[45] amongst others, have applied Cogswell's analysis with minor modifictions. In a series of papers on the subject,[76,11,77,78] the present author has drawn heavily on the original ideas of Cogswell in the belief that the fundamental philosophy is sound. As a result it has been possible to build up a picture of contraction flows that reflects qualitatively, and often quantitatively, most, if not all, of the various phenomena that have been described above.

First we consider some general aspects of entry flow in order to define our notation and certain parameters, although we restrict consideration to the axisymmetric case. The approach taken here is identical to that adopted by Vrentas and Vrentas.[39]

Referring to Fig. 1.1 we apply the balance principles in a global form to the volume V of fluid enclosed by the surface S which is comprised of the walls of the contraction and the entry and exit planes S_1 and S_2, respectively. These planes are assumed to be located in regions of fully developed Poiseuille flow. Then, conservation of mass may be written as

$$Q = \int_{S_1} -\mathbf{v} \cdot \mathbf{n} \, dS = \int_{S_2} \mathbf{v} \cdot \mathbf{n} \, dS \tag{1.1}$$

where Q is the volumetric flow rate, $\mathbf{v}$ is the velocity field and $\mathbf{n}$ is the usual outward normal to the surface element dS.

The total stress tensor $\mathbf{T}$ is expressed as

$$\mathbf{T} = -p\mathbf{I} + \boldsymbol{\sigma} \tag{1.2}$$

where p is an isotropic pressure distribution and $\boldsymbol{\sigma}$ is an extra-stress tensor.

Conservation of momentum is first written in its spatial representation as

$$\text{div } \mathbf{T} + \rho\mathbf{b} = \rho\mathbf{a} \tag{1.3}$$

where ρ is the fluid density, $\mathbf{b}$ is the body force per unit mass and $\mathbf{a}$ is the acceleration. Multiplying (1.3) scalarly by $\mathbf{v}$ and integrating over the volume V yields the following mechanical energy balance:

$$\int_S -p(\mathbf{v} \cdot \mathbf{n}) \, dS + \int_S \mathbf{v} \cdot \boldsymbol{\sigma} \cdot \mathbf{n} \, dS + \int_S \rho(\mathbf{v} \cdot \mathbf{v})(\mathbf{v} \cdot \mathbf{n}) \, dS$$

$$= \int_V \boldsymbol{\sigma} : \mathbf{d} \, dV - \int_V \mathbf{n} \cdot \mathbf{b} \cdot \mathbf{v} \, dS \tag{1.4}$$

where $\mathbf{d} = (\mathbf{L} + \mathbf{L}^T)/2$ is the rate-of-strain tensor, $\mathbf{L}$ being the velocity gradient. In deriving this equation, steady flow and fluid incompressibility have been assumed. The left-hand side of (1.4) represents the total rate of working of the stresses on the surface S plus a term corresponding to the net increase in kinetic energy. The right-hand side contains the total rate of dissipation of stress with a contribution due to the body forces.

Clearly, for a given constitutive equation, combined with the relevant boundary conditions, eqns (1.1) and (1.3) are, in principle, sufficient to determine fully the velocity field. However, the mechanical energy balance given by eqn (1.4) is simply a necessary, but clearly insufficient, condition for the momentum balance eqn (1.3) to be satisfied and cannot be used in conjunction with the latter. On the other hand, if the mechanical energy balance is to be used in place of the momentum balance, then additional necessary information is required. That additional information will be expressed in the form of the following postulate that:

of all the steady, isochoric velocity fields satisfying eqns (1.1) and (1.4), the one adopted by the fluid in order to satisfy conservation of momentum is the one that minimises the total rate of working of the pressure field.

This may be viewed by some as a controversial statement. Denn[79] has discussed the use of this principle concluding that, 'although unproven, it is nonetheless physically plausible'. In fact, a principle of this type represents a restriction on the form of the stress constitutive relation and, as such, cannot be proven in general. Whether or not such a restriction is generally valid for real materials remains a matter for debate.

We assume no-slip velocity boundary conditions on the geometry wall and neglect the effect of body forces. Since the velocity is then known everywhere on the boundary and the extra stresses are known everywhere on the boundary where the velocity is non-zero, the above postulate is seen to be equivalent to assuming that the velocity field minimises the total rate of dissipation within the volume V

It is not difficult to show[39] that in the fully developed upstream and downstream regions, the pressure distribution takes on the form:

$$p(r, z) - \sigma_{zz} = T - N_1 - \int_r^R N_2/r' \, \mathrm{d}r' \tag{1.5}$$

where T is the total radial normal stress on the wall of the geometry, as would be measured by a pressure transducer, and R is the radius of the corresponding cylinder. N_1 and N_2 are the usual first and second normal stress differences in shear and are both functions of the local shear rate. Notice that if N_2 is zero, then the centre-line pressure is equal to the measured wall stress T, since we can assume without loss of generality that all extra-stresses are zero on the axis of symmetry, in the case of fully developed flow.

It is not necessary, at this stage, to neglect the effect of N_2. Certainly it is known to be small relative to N_1 and also of opposite sign, but it is not immediately clear that it can be neglected in the context of, for example, eqn (1.5). On the other hand, N_2 is notoriously difficult to measure accurately for the majority of materials. Also, one of the principal points of the current exercise is to obtain an understanding of the physical mechanisms involved in contraction flows. We do not want to obscure the picture by carrying forward equations that are excessively complicated. An inspection of eqn (1.5) indicates that the effect of N_2 may be equivalent to a lowering of the value of N_1. Therefore, more for convenience than for any other reason, N_2 will be assumed to be zero.

The mechanical energy balance eqn (1.4) can now be written as

$$(T_1 - T_2)Q = (p_1 - p_2)_{r=0}Q$$

$$= \int_V \boldsymbol{\sigma} : \mathbf{d} \, dV - 2\pi \int_0^{R_2} (N_1 \mathbf{v} . \mathbf{n})_2 r \, dr + 2\pi \int_0^{R_1} (N_1 \mathbf{v} . \mathbf{n})_1 r \, dr$$

$$(1.6)$$

where the suffixes 1 and 2 refer to the inlet and outlet planes, respectively.

The entry pressure drop is often quoted in non-dimensional form by means of the Couette correction C, defined by

$$C = (p_1 - p_2 - p_{fd})_{r=0}/(2\sigma_w) \tag{1.7}$$

where p_{fd} is the pressure drop that would be expected on the basis of fully developed Poiseuille flow and σ_w is the fully developed shear stress at the downstream capillary wall.

In practice, the second surface integral in eqn (1.6) is usually negligible because the upstream diameter is generally very much larger than the downstream diameter, leading to very small extra-stress levels at the inlet plane. However, the first surface integral cannot be neglected. In fact, it is this term that accounts for the reduction in the value of the Couette correction that is observed at low strain rates in the majority of numerical simulations.

1.4.2. Approximating the Stress Field

Because we wish to investigate the influence of various rheometrical properties on the characteristics of contraction flows, we do not want to invoke a specific stress constitutive equation for use with the mechanical energy balance. Instead, we will replace the stress tensor by an approximation involving the aforementioned rheometric functions. First we will make an attempt at justifying the procedure and, at the same time, make some comments regarding what is meant by 'elasticity'.

In a steady simple shear flow, with velocity components

$$v_1 = \dot{\gamma} x_2, \qquad v_2 = v_3 = 0 \tag{1.8}$$

when referred to a Cartesian coordinate system (x_1, x_2, x_3), the corresponding components of the rate-of-strain and extra-stress tensors may be written as

$$\mathbf{d} = \begin{vmatrix} 0 & \dot{\gamma}/2 & 0 \\ \dot{\gamma}/2 & 0 & 0 \\ 0 & 0 & 0 \end{vmatrix}, \qquad \boldsymbol{\sigma} = \begin{vmatrix} N_1 + N_2 & \sigma & 0 \\ \sigma & N_2 & 0 \\ 0 & 0 & 0 \end{vmatrix} \tag{1.9}$$

where σ, the shear stress, and N_1 and N_2, the first and second normal stress differences, are functions of the constant shear rate $\dot{\gamma}$.

It will be noted that the stress power, or, alternatively, the rate of dissipation per unit volume is given by

$$\boldsymbol{\sigma} : \mathbf{d} = \dot{\gamma}\sigma \tag{1.10}$$

This is clearly independent of the normal stress differences N_1 and N_2 which may, therefore, be regarded as being of purely elastic origin. The shear stress, however, is not purely a viscous response, since it may contain a recoverable component. Nevertheless, it is indistinguishable from a purely viscous response.

Similarly, a pure steady uniaxial extensional flow is characterised by the following velocity distribution:

$$v_1 = \dot{\varepsilon} x_1, \qquad v_2 = -\dot{\varepsilon} x_2/2, \qquad v_3 = -\dot{\varepsilon} x_3/2 \tag{1.11}$$

The components of the rate-of-strain and extra-stress tensors for this flow can be expressed as

$$\mathbf{d} = \begin{vmatrix} \dot{\varepsilon} & 0 & 0 \\ 0 & -\dot{\varepsilon}/2 & 0 \\ 0 & 0 & \dot{\varepsilon}/2 \end{vmatrix}, \qquad \boldsymbol{\sigma} = \begin{vmatrix} \sigma_E & 0 & 0 \\ 0 & 0 & 0 \\ 0 & 0 & 0 \end{vmatrix} \tag{1.12}$$

where the tensile stress σ_E is in general a function of the constant stretch rate $\dot{\varepsilon}$.

For this flow, the stress power is given by

$$\boldsymbol{\sigma} : \mathbf{d} = \dot{\varepsilon}\sigma_E \tag{1.13}$$

Again, the tensile stress is a manifestation of both viscous and elastic effects, but those two effects cannot be separated and it is therefore indistinguishable from a purely viscous response. Unlike shear flow, extensional flow does not exhibit what might be regarded as a purely elastic response.

In view of the above observations, and in the context of the current discussion only, the term 'elasticity' will be associated with the existence of non-zero normal stress differences in shear flow.

Consider now the superposition of the two velocity fields described by eqns (1.8) and (1.11). The rate-of-strain tensor is additive and for the combined flow can be written as

$$\mathbf{d} = \begin{vmatrix} \dot{\varepsilon} & \dot{\gamma}/2 & 0 \\ \dot{\gamma}/2 & -\dot{\varepsilon}/2 & 0 \\ 0 & 0 & -\dot{\varepsilon}/2 \end{vmatrix} \tag{1.14}$$

The stress tensor, however, is not additive. Nevertheless, by assuming additivity of stresses, the resulting expression is exact in the two extremes of simple shear flow and pure extensional flow. As a point of interest, it also turns out to be exact in the case of a second-order fluid. Also, there is some numerical evidence to suggest that such additivity may indeed be approximately valid.[56] Therefore,

$$\boldsymbol{\sigma} = \begin{vmatrix} \sigma_E + N_1 + N_2 & \sigma & 0 \\ \sigma & N_2 & 0 \\ 0 & 0 & 0 \end{vmatrix} \tag{1.15}$$

may be considered as an approximation of the stress field corresponding to the strain-rate field (eqn (1.14)).

The above discussion illustrates the philosophy that will be applied to the contraction flow problem and also exemplifies the seriousness of the assumptions and approximations that are necessary.

Returning to the problem at hand, recall that the experimental evidence depicted schematically in Fig. 1.3 suggests the existence of two distinct modes of flow. At low strain rates the flow just upstream of the contraction plane is essentially radial and does not detach from the geometry wall at the lip of the contraction. The second mode involves flow through a stream tube that detaches from the geometry at the lip of the contraction with re-attachment occurring some distance upstream of the contraction plane. These two modes will be dealt with separately and the way in which the fluid changes from one mode to the other, as well as the reason for the change, will be discussed in the light of the results obtained.

1.4.3. Quasi-radial Flow

Consider the flow field to be approximated as shown in Fig. 1.4. Define the usual spherical coordinate system (r, θ, ξ) with origin 0 on the axis of symmetry, in the contraction plane. Define a radial flow with velocity components given by

$$v_r = -Qf(\theta)/2\pi r^2, \qquad v_\theta = v_\xi = 0 \tag{1.16}$$

This velocity distribution is assumed to hold upstream of the contraction plane and for $R_2 < r < R_u$. R_2 is the radius of the downstream capillary and it will be assumed that $R_u \gg R_2$. The function $f(\theta)$ has been normalised such that

$$\int_0^{\pi/2} f(\theta) \sin(\theta) \, d\theta = 1 \tag{1.17}$$

in order that conservation of mass be satisfied automatically.

The components of the rate-of-strain tensor may be written as

$$\mathbf{d} = \begin{vmatrix} \dot{\varepsilon} & \dot{\gamma}/2 & 0 \\ \dot{\gamma}/2 & -\dot{\varepsilon}/2 & 0 \\ 0 & 0 & -\dot{\varepsilon}/2 \end{vmatrix} \tag{1.18}$$

where

$$\dot{\varepsilon} = Qf(\theta)/\pi r^3 \quad \text{and} \quad \dot{\gamma} = -Qf'(\theta)/2\pi r^3 \tag{1.19}$$

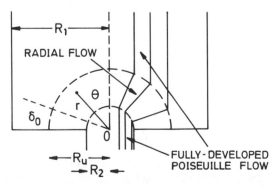

Fig. 1.4. Schematic description of quasi-radial flow.

will be regarded as local stretch and shear rates, respectively. In general, the components of the extra-stress tensor take the form

$$\boldsymbol{\sigma} = \begin{vmatrix} \sigma_{rr} & \sigma_{r\theta} & 0 \\ \sigma_{r\theta} & \sigma_{\theta\theta} & 0 \\ 0 & 0 & \sigma_{\xi\xi} \end{vmatrix} \tag{1.20}$$

Hence, the rate of dissipation per unit volume becomes

$$\boldsymbol{\sigma} : \mathbf{d} = (\sigma_{rr} - \sigma_{\theta\theta})\dot{\varepsilon} + \sigma_{r\theta}\dot{\gamma} + (\sigma_{\theta\theta} - \sigma_{\xi\xi})\dot{\varepsilon}/2 \tag{1.21}$$

We now replace the stresses in eqn (1.21) in terms of fundamental rheometric properties, the pros and cons of which have been amply discussed above and in Ref. 56. Equation (1.21) therefore becomes

$$\boldsymbol{\sigma} : \mathbf{d} = (\sigma_E + N_1)\dot{\varepsilon} + \sigma\dot{\gamma} + N_2\dot{\varepsilon}/2 \tag{1.22}$$

In order to make progress analytically, N_2 is assumed to be zero and power-law representations are used for each of the other rheometric functions:

$$\sigma = \dot{\gamma}\eta = k\dot{\gamma}^n$$
$$\sigma_E = \dot{\varepsilon}\eta_E = l\dot{\varepsilon}^t \tag{1.23}$$
$$N_1 = m\dot{\gamma}^{p+1}$$

where η and η_E are the shear and extensional viscosities, respectively.
   Combining eqns (1.22) and (1.23) yields

$$\int_V \boldsymbol{\sigma} : \mathbf{d} \, dV = (2Q/3) \int_0^{\pi/2} G(f, f') \sin(\theta) \, d\theta + \dot{E}_h + \dot{E}_{cap} \tag{1.24}$$

where

$$G = (l/t)A'f'^{t+1} + (k/n)A''(-f'/2)^{n+1} + mA^{p+1}f(-f'/2)^{p+1}/(p+1) \tag{1.25}$$

and

$$A = Q/\pi R_2^3$$

$\dot{E}_h$ is a term added to account for the contribution to the rate of dissipation arising from the fluid contained in the hemispherical region $r < R_2$. If that region is assumed to be simply an extension of the downstream capillary in which the flow is fully developed Poiseuille flow, then it can be shown that

$$\dot{E}_{\mathrm{h}} = 2kQ \left(\frac{3n+1}{n}\right)^{n+1} A^n \int_0^1 \phi^{(n+1)/n}(1-\phi^2)^{\frac{1}{2}}\phi \, \mathrm{d}\phi \tag{1.26}$$

Also, $\dot{E}_{\mathrm{cap}}$ is the contribution due to the fully developed flow in the downstream capillary. This is easily evaluated for power-law fluids but it will not be required.

Variational calculus may be employed to minimise the total rate of dissipation given by eqn (1.24) with respect to the function $f(\theta)$, subject to the constraint expressed by eqn (1.17). However, the resulting differential equation is not particularly enlightening for current purposes. Instead, we shall approximate $f(\theta)$ to the following form:

$$f(\theta) = \begin{cases} f_0, & \text{for } 0 < \theta < \pi/2 - \delta_0 \\ f_0((\pi/2 - \theta)/\delta_0, & \text{for } \pi/2 - \delta_0 < \theta < \pi/2 \end{cases} \tag{1.27}$$

where δ_0 is assumed to be small. With this form for $f(\theta)$, shearing is assumed to be restricted to a thin, circular, wedge-shaped 'boundary layer' of 'thickness' δ_0 adjacent to the contraction plane.

In addition we will restrict attention to fluids for which $p = n$. This equality is satisfied for so-called Boger-type fluids and, indeed, all fluids in the limit of low strain rates. Since we have already anticipated that this flow mode will be preferred at low strain rates, this restriction is not a particularly serious one. In general, however, the contribution of elasticity will be over-estimated.

Substituting eqn (1.27) into (1.24), keeping only the leading terms in δ_0 it can be shown after some straightforward algebra, that for minimum rate of dissipation the 'boundary-layer thickness' is given by

$$\delta_0^{n+1} = k(t+2)A^{n-t}\left\{1 + \frac{mnA}{2k(n+1)}\right\}\bigg/ 2^n l(t+1) \tag{1.28}$$

This equation is very illuminating since it is able to indicate the way in which the streamlines adjust themselves, depending on the relative importance of each of the various rheometrical parameters.

First, for consistency, it is required that δ_0 be small. Clearly, from eqn (1.28), provided that $t > n+1$ (which is not unusual for polymer solutions), it is always possible to choose sufficiently high flow rates for δ_0 to become as small as necessary. Indeed, under those circumstances the 'boundary-layer thickness' decreases with increasing flow-rate. That is, the streamlines are pushed towards the contraction plane. Increased shear-thinning (smaller n) and increased stretch-hardening (larger t) enhance this behaviour. Although we cannot strictly associate δ_0 with the size of the salient corner vortex, it would seem logical to assume that a reduced value of δ_0 indicates that such a vortex would be reduced in size. This is consistent with the 'bulb-flow' behaviour depicted in Fig. 1.3(d) and also in general terms with the situation that occurs immediately before the appearance of a lip-vortex.

Increasing the extensional viscosity level (l) reduces the 'boundary-layer thickness' whereas increasing the shear viscosity (k) or elasticity (m) has the opposite effect, resulting in a thicker layer. Only for Newtonian fluids does δ_0 become independent of flow rate and in that case $\delta_0 = 1/2$. Thus we have a mechanism that is consistent in general terms with experimental observations of the growth or decay of the salient corner vortex.

Substitution of eqns (1.28) and (1.27) into eqn (1.24) and then into eqns (1.6) and (1.7) yields the following expression for the Couette correction:

$$C = \alpha_1 A^{t-n} + \alpha_2 (1 + \alpha_3 A)^{1/(1+n)} A^{n(t-n)/(1+n)} - \alpha_4 A + C_h \qquad (1.29)$$

where

$$C_h = \{(3n+1)/n)\} \int_0^1 \phi^{(1+n)/n}(1-\phi^2)^{\frac{1}{2}} \phi \, d\phi$$

$$\alpha_1 = l\{n/(3n+1)\}^n/3kt$$

$$\alpha_2 = \frac{n+1}{3kn}\left(\frac{n}{3n+1}\right)^n \left\{\left(\frac{t+1}{t+2}\right)^n \frac{l^n k}{2^{2n+1}}\right\}^{1/(1+n)}$$

$$\alpha_3 = mn/2k(1+n)$$

and

$$\alpha_4 = m(3n+1)/2k(2n+1)$$

We shall return to the Couette correction later, but before leaving this section it is worth noting the term involving α_4 in eqn (1.29) which arises essentially from the downstream fully developed stress boundary condition. For parameter values of practical interest, it is that term which causes an initial reduction in the value of C, before being dominated by the other positive terms at higher flow rates. In other words, the numerically observed reduction in C is a consequence of the way in which it is defined, combined with the influence of the downstream stress boundary conditions.

1.4.4. Funnel Flow

The second flow mode is one that was the theme of an earlier study by the present author[76] in which the same analytical technique was employed. The flow in the upstream region was assumed to be locally fully developed Poiseuille flow through a slowly converging funnel which essentially presupposes the existence of a large recirculating region extending up to the lip of the contraction (similar to Fig. 1.3(c), for example). Elasticity was not included in the original study, although in a later paper,[70] the analysis was modified to take account of elasticity *but only in as much as it affected the downstream stress boundary conditions.*

Elasticity may be included in that analysis in exactly the same way that it was treated in the above quasi-radial flow. In some ways the funnel-flow analysis is more pleasing, since the minimisation of the rate of dissipation can be performed using variational analysis, with respect to a whole class of velocity fields, without the need to make an approximation of the type used in eqn (1.27). It is also more amenable for use as a means of estimating the extensional viscosity from measurements of the entry pressure drop.

Referring to Fig. 1.5, the flow is assumed to be confined to a funnel of radius $R(z)$ where z is the axial direction of a cylindrical coordinate system (r, θ, z) with the origin again on the axis of symmetry and in the plane of the contraction. The same power-law expressions (eqn (1.23)) will again be employed.

The axial velocity component at any given value of z is assumed to be that corresponding to fully developed Poiseuille flow in a circular cylinder of radius $R(z)$:

$$v_z = \frac{(3n+1)Q}{(n+1)\pi R^2} \left\{ 1 - \left(\frac{r}{R}\right)^{1+1/n} \right\} \tag{1.30}$$

Conservation of mass is then invoked to determine the corresponding radial component

$$v_r = \frac{(3n+1)Qr}{(n+1)\pi R^3} \left\{ 1 - \left(\frac{r}{R}\right)^{1+1/n} \right\} \frac{\mathrm{d}R}{\mathrm{d}z} \tag{1.31}$$

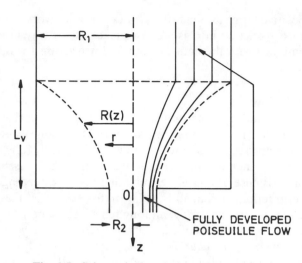

Fig. 1.5. Schematic description of funnel flow.

The non-zero components of the rate-of-strain tensor are then

$$d_{rr} = \frac{(3n+1)Q}{(n+1)\pi R^3} \left\{ 1 - \left(\frac{2n+1}{n}\right)\left(\frac{r}{R}\right)^{1+1/n} \right\} \frac{dR}{dz}$$

$$d_{\theta\theta} = \frac{(3n+1)Q}{(n+1)\pi R^3} \left\{ 1 - \left(\frac{r}{R}\right)^{1+1/n} \right\} \frac{dR}{dz}$$

$$d_{zz} = \frac{(3n+1)Q}{(n+1)\pi R^3} \left\{ \left(\frac{3n+1}{n}\right)\left(\frac{r}{R}\right)^{1+1/n} - 2 \right\} \frac{dR}{dz} \qquad (1.32)$$

$$d_{rz} = -\frac{(3n+1)Q}{2n\pi R^3} \left(\frac{r}{R}\right)^{1/n}$$

where terms involving $(dR/dz)^2$ and d^2R/dz^2 have been neglected. That is, the convergence angle is assumed to be both small and slowly varying.

Following the example of the previous section, the total rate of dissipation within the funnel and capillary is given by

$$\int_V \boldsymbol{\sigma} : \mathbf{d}\, dV = \int_0^{-L_v} F(R, dR/dz)\, dz + \dot{E}_{cap} \qquad (1.33)$$

where

$$F = (l\dot{\varepsilon}^t + m\dot{\gamma}^p)\dot{\varepsilon} + k\dot{\gamma}^n$$

with

$$\dot{\varepsilon} = d_{zz} \quad \text{and} \quad \dot{\gamma} = 2d_{rz}$$

and again the second normal stress difference in shear, N_2, has been assumed to be zero.

The rate of dissipation given by eqn (1.33) can be minimised with respect to all smooth profiles, $R(z)$, using standard variational calculus. The resulting optimum profile is given by

$$\left(-\frac{dR}{dz}\right)^{t+1} = \frac{k(n+1)^{t+1}}{lt(3n+1)n^n I_{nt}} \left\{\frac{(3n+1)Q}{\pi R^3}\right\}^{n-t}$$

where

$$I_{nt} = \int_0^1 \left[\text{abs}\left\{2 - \left(\frac{3n+1}{n}\right)\phi^{1+1/n}\right\}\right]^{t+1} \phi \, d\phi \qquad (1.34)$$

The first important point to be noticed about eqn (1.34) is that the predicted profile of the funnel is independent of elasticity, i.e. the first normal stress difference in shear. Also, eqn (1.34) is readily integrated to yield an expression for the vortex length ratio. The relevant equations are to be found in Ref. 76. It is easily deduced that the vortex length increases with increasing flow rate whenever $t > n$. That is, vortex enhancement is expected to occur whenever the Trouton ratio is an increasing function of strain rate.

Note also that straight-edged vortices, independent of flow rate, are predicted if the Trouton ratio is constant $(t = n)$, in which case the convergence angle decreases as the actual Trouton value increases. This result ties in very well with the analyses of Batchelor[65] and Goddard[66] for the extensional viscosity of fibre-filled fluids and the experimental observations of Binnington and Boger.[12] These latter comments clearly do not apply to Newtonian fluids for which the Trouton ratio has the constant value of 3. The reason for this will become clear later.

Although the vortex length appears to be independent of elasticity, the pressure drop is not. The resulting expression for the Couette correction is†

$$C = \beta_1 A^{(t-n)/(1+t)} - \beta_2 A^{p+1-n} \qquad (1.35)$$

† It should be noted that the restriction $p = n$ is not necessary to obtain analytical expressions for this flow and so that restriction has not been made here.

where

$$\beta_1 = \frac{(1+t)^2}{3t^2(1+n)^2} \left\{ \frac{lt(3n+1)^{1+t-n}n^n I_{nt}}{k} \right\}^{1/(1+t)}$$

and

$$\beta_2 = m(3n-1)(3n+1)\{(3n+1)/n\}^{p+1-n}/\{3k(2n+p+1)(3n+p+2)\}$$

Again, eqn (1.35) indicates that at low flow rates, elasticity has the effect of reducing the value of the Couette correction. Indeed, only in the case of severe shear thinning ($n < 1/3$) is elasticity predicted to contribute positively to C. It is pertinent to point out also that elasticity may become particularly significant in the case of Boger fluids for which $p = n = 1$.

1.4.5. Discussion of Analytical Results

The two flow modes described in the previous sections appear to reflect the various observations that have been made in visualisation studies. It remains now to deduce why the fluid adopts a particular mode under specific conditions. To do this we must look more closely at the two expressions for the Couette correction.

First, for Newtonian fluids the predicted values for the Couette correction are 0·87 and 0·94 for the quasi-radial and funnel flows, respectively. These compare with the generally accepted numerical result of Christiansen *et al.*[80] of 0·69. As might be expected, the true value is somewhat lower than the approximate ones. Nevertheless, the point remains that the quasi-radial flow yields a lower value than the funnel flow, and so the former will be the preferred mode at all flow rates. Also, since all fluids respond as Newtonian ones at sufficiently low strain rates, it follows that the radial-type flow is the preferred mode for all fluids in the limit of low flow rates.

For non-Newtonian fluids, the departure from radial flow may occur in two ways, as discussed earlier. Depending on whether elasticity or extensional viscosity is dominant at low flow rates, the salient corner vortex will grow or decay, respectively. At the same time the Couette correction will be increased by extensional effects but reduced by elasticity. For realistic parameters, a net reduction is to be expected.

For typical polymer solutions, the extensional viscosity is a rapidly increasing function of stretch rate ($t > 1$). In this case the streamlines of the radial flow are pushed towards the contraction plane, but there is clearly a limit to this process. Eventually, the strain rates increase in

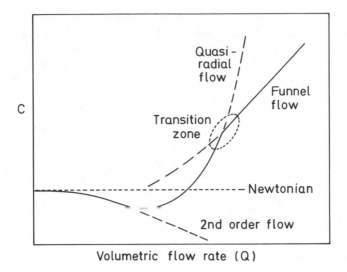

Fig. 1.6. Typical form of a Couette correction *versus* flow-rate plot.

proportion to the flow rate, leading to stresses dominated by extensional viscosity and, therefore, exceptionally high values of the Couette correction ($\sim Q'$). However, in funnel flow the rate of increase of the stresses is lower on account of the fact that the funnel can continuously be enlarged in order to reduce the rate at which tensile stresses grow with flow rate. In this case, $C \sim Q'^{-n}$. There will be, therefore, a critical flow rate at which funnel flow becomes preferred. The maximum stretch rate is predicted to occur at the lip of the contraction[76] and so it is there that departure from radial flow will first take place. Hence, the lip vortex can be seen to be the initiation of the build-up of the funnel.

Other scenarios are possible, including the smooth transition of a growing salient corner vortex into funnel flow. However, because of the number of independent rheometric parameters involved, it is not practical to analyse every possible situation.

On the basis of eqns (1.29) and (1.35), the Couette correction will have the form shown schematically in Fig. 1.6, for realistic material parameters.

1.5. ESTIMATING EXTENSIONAL VISCOSITY

Given that the analysis is able to reflect the many general features that are observed in contraction flow, it is reasonable to assume that the effect of extensional viscosity has been accounted for in a physically realistic manner. It is, therefore, appropriate to make use of the analysis in order to extract, from experimental pressure-drop data, what we might call an extensional viscosity. It must be borne in mind that such a quantity will not be, indeed cannot be, the precise extensional viscosity as defined for a pure extensional flow. Nevertheless, it will be a quantity that closely reflects the extensional behaviour of the fluid.

Experimentally, the procedure is straightforward. Pressure drops are measured as a function off low rate for two different capillary lengths. The data are then extrapolated, for each flow rate, to obtain the pressure drop corresponding to zero capillary length. This, then, is the entry pressure drop. Alternatively, a 'nominally zero-length' capillary may be used, since the analysis does not distinguish between flow through a capillary and flow through an orifice. The Couette correction is then easily calculated.

To obtain an estimate for the extensional viscosity, assume funnel flow, since the dependence of C on elasticity for this flow is manageable. From eqn (1.35) (second term on left-hand side) the contribution due to elasticity can be evaluated and added to C to yield a modified Couette correction. An inspection of eqn (1.35) shows that a plot of this modified correction *versus* flow rate, on logarithmic scales, has a gradient of $(t - n)/(1 + t)$, from which the value of t can be determined. A simple application of eqn (1.35) at one flow rate then yields the value of l. It is, of course, necessary to have determined beforehand the four parameters associated with the shear viscosity and first normal stress difference N_1. The approximate range of stretch rates for which the power-law parameters l and t are valid can be obtained by deducing the maximum stretch rate in the flow. Details of this can be found in Ref. 76.

The technique has been applied to several polymeric fluids and also to polymer melts and composites. In the case of highly shear-thinning polymer solutions, the results compare remarkably well with measurements obtained from the 'spin-line rheometer.'[11] For the fluids studied elasticity had little influence on the Couette correction. A somewhat different conclusion must be drawn when highly elastic 'Boger-type' fluids are studied,[11,78] since now the contraction analysis yields an extensional viscosity that is lower than that obtained from the spin-line technique. This should come as no surprise because of the long relaxation times exhibited by the fluids and the limited residence times pertinent to the flow. In other words, the estimated extensional viscosity must be associated with a stress field that is transient, in the Lagrangian sense.

Further evidence of the usefulness of the above analytical approach is provided by its ability to explain anomalous data from a lubricated converging-flow rheometer.[77]

Estimated extensional viscosity data from the contraction flow of glass-fibre-filled polypropylene composites[81] have been shown to be consistent with the theoretical predictions of Bachelor[65] and Goddard[66] for suspensions of inextensible fibres. Indeed, these analyses may be used to provide a limiting high strain rate limit for the extensional viscosity of polymeric solutions, by assuming that a fully extended polymer chain behaves as an inextensible fibre under such conditions. Conversely, by measuring the limiting extensional viscosity, an estimate of the 'fibre' length/diameter ratio (and hence molecular weight) can be obtained. An acceptable value of the molecular weight of the A1 test fluid was obtained in this way,[82] thus providing further support for the underlying analytical approach.

1.6. CONCLUDING REMARKS

There is little doubt now that extensional viscosity has a major influence on the way in which fluids flow through contractions and similar geometries. The analysis presented in this chapter represents an attempt to demonstrate how the extensional viscosity, as well as other rheometric properties, might be expected to influence various aspects of the flow. It is by necessity a very approximate analysis, guided as much, if not more, by physical intuition as it is by mathematical rigour.

The fact that so many of the observed flow features are reflected in the analytical results is a good indication that the fluids' properties have been accounted for in a physically realistic way. Nevertheless, much work needs to be done, particularly by the numerical analysts, before the value of the various assumptions can be quantified.

Even before such validation, it is clear that contraction flows are capable of distinguishing between fluids that differ only in their response to extensional flows. Hence, an extensional viscosity based on contraction flow must be regarded as meaningful.

Finally, it should be remembered that this chapter represents the personal view that the author has of contraction flows. During the development of the chapter some digressions have been made in order to introduce what might be regarded as circumstantial evidence. This has been done deliberately with the aim of stimulating further discussion of a topic that has not been exhausted.

REFERENCES

1. S. A. White, A. D. Gotsis and D. G. Baird, *J. Non-Newtonian Fluid Mech.*, 1987, **24**, 121.
2. D. V. Boger, *Ann. Rev. Fluid Mech.*, 1987, **19**, 157.
3. D. V. Boger and M. M. Denn, *J. Non-Newtonian Fluid Mech.*, 1980, **6**, 163.
4. K. Walters, *Rheometry*, Chapman & Hall, London, 1976.
5. D. V. Boger, D. U. Hur and R. J. Binnington, *J. Non-Newtonian Fluid Mech.*, 1986, **20**, 31.
6. M. E. Kim-E, R. A. Brown and R. C. Armstrong, *J. Non-Newtonian Fluid Mech.*, 1983, **13**, 341.
7. K. Chiba, T. Sakatani and K. Nakamura, *J. Non-Newtonian Fluid Mech.* 1990, **36**, 193.
8. R. E. Evans and K. Walters, *J. Non-Newtonian Fluid Mech.*, 1989, **32**, 95.
9. R. E. Evans and K. Walters, *J. Non-Newtonian Fluid Mech.*, 1986, **20**, 11.
10. D. M. Binding, K. Walters, J. Dheur and M. J. Crochet, *Phil. Trans. Roy. Soc. Lond.*, *A*, 1987, **323**, 449.
11. D. M. Binding and K. Walters, *J. Non-Newtonian Fluid Mech.*, 1988, **30**, 233.
12. R. J. Binnington and D. V. Boger, *J. Non-Newtonian Fluid Mech.*, 1987, **26**, 115.
13. G. G. Lipscomb II, M. M. Denn, D. U. Hur and D. V. Boger, *J. Non-Newtonian Fluid Mech.*, 1988, **26**, 297.
14. D. V. Boger and R. J. Binnington, *J. Non-Newtonian Fluid Mech.*, 1990, **35**, 339.
15. J. M. Piau, N. El Kissi and B. Tremblay, *J. Non-Newtonian Fluid Mech.*, 1988, **30**, 197.
16. E. B. Bagley and A. M. Birks, *J. Appl. Phys.*, 1960, **31**, 556.
17. T. F. Ballenger, I. J. Chen, J. W. Crowder, G. E. Hagler, D. C. Bogue and J. L. White, *Trans. Soc. Rheol.*, 1971, **15**, 195.
18. J. L. White and A. Kondo, *J. Non-Newtonian Fluid Mech.*, 1978, **3**, 41.
19. C.-Y. Ma, J. L. White, F. C. Weissert and K. Min, *J. Non-Newtonian Fluid Mech.*, 1985, **17**, 275.
20. S. A. White and D. G. Baird, *J. Non-Newtonian Fluid Mech.*, 1986, **20**, 93.
21. S. A. White and D. G. Baird, *J. Non-Newtonian Fluid Mech.*, 1988, **29**, 245.
22. H. Nguyen and D. V. Boger, *J. Non-Newtonian Fluid Mech.*, 1979, **5**, 353.
23. G. Dembek, *Rheol. Acta*, 1982, **21**, 553.
24. K. Walters and D. M. Rawlinson, *Rheol. Acta*, 1982, **21**, 547.
25. K. Walters and M. F. Webster, *Phil. Trans. Roy. Soc. Lond.*, *A*, 1982, **308**, 199.
26. A. B. Metzner, E. A. Uebler and C. F. Chan Man Fong, *AIChE J.*, 1969, **15**, 750.
27. R. J. Binnington, G. J. Troup and D. V. Boger, *J. Non-Newtonian Fluid Mech.*, 1983, **12**, 255.

28. Y. Xu, P. Wang and R. Qian, *Rheol. Acta*, 1986, **25**, 239.
29. T. Hasegawa and T. Iwaida, *J. Non-Newtonian Fluid Mech.*, 1984, **15**, 279.
30. J. V. Lawler, S. J. Muller, R. A. Brown and R. C. Armstrong, *J. Non-Newtonian Fluid Mech.*, 1986, **20**, 51.
31. W. P. Raiford, L. M. Quinzani, P. J. Coates, R. C. Armstrong and R. A. Brown, *J. Non-Newtonian Fluid Mech.*, 1989, **32**, 39.
32. A. M. Wunderlich, P. O. Brunn and F. Durst, *J. Non-Newtonian Fluid Mech.*, 1988, **28**, 267.
33. M. Shirakashi and M. Watanabe, *Proc. Xth Int. Congr. Rheol.*, 1988, **2**, 269.
34. R. L. Boles, H. L. Davis and D. C. Bogue, *Polym. Eng. Sci.*, 1970, **10**, 24.
35. G. D. Eisenbrand and J. D. Goddard, *J. Non-Newtonian Fluid Mech.*, 1982, **11**, 37.
36. J. L. Duda and J. S. Vrentas, *Trans. Soc. Rheol.*, 1973, **17**, 89.
37. M. Moan, G. Chauveteau and S. Ghoniem, *J. Non-Newtonian Fluid Mech.*, 1979, **5**, 463.
38. J. S. Vrentas, J. L. Duda and S.-A. Hong, *J. Rheol.*, 1982, **26**, 347.
39. J. S. Vrentas and C. M. Vrentas, *J. Non-Newtonian Fluid Mech.*, 1983, **12**, 211.
40. D. H. Crater and J. A. Cucol, *J. Polym. Sci.*, 1984, **22**, 1.
41. T. Hasegawa and T. Iwaida, *J. Non-Newtonian Fluid Mech.*, 1984, **15**, 257.
42. G. Chauveteau, M. Moan and A. Magueur, *J. Non-Newtonian Fluid Mech.*, 1984, **16**, 315.
43. A. G. Gibson and G. A. Williamson, *Polym. Eng. Sci.*, 1985, **25**, 968.
44. A. G. Gibson and G. A. Williamson, *Polym. Eng. Sci.*, 1985, **25**, 980.
45. T. H. Kwon, S. E. Shen and K. K. Wang, *Polym. Eng. Sci.*, 1986, **26**, 214.
46. M. Tachibana, N. Kawabata and H. Genno, *J. Rheol.*, 1986, **30**, 517.
47. C. A. Heiber, *Rheol. Acta*, 1987, **26**, 92.
48. T. Hasegawa and F. Fukutomo, *Proc. Xth Int. Congr. Rheol.*, 1988, **1**, 395.
49. M. J. Crochet and K. Walters, *Ann. Rev. Fluid Mech.*, 1983, **15**, 241.
50. M. J. Crochet, A. R. Davies and K. Walters, *Numerical Simulation of Non-Newtonian Flow*, Elsevier, Amsterdam, 1984.
51. L. G. Leal, M. M. Denn and R. Keunings (guest eds), *J. Non-Newtonian Fluid Mech.*, 1988, Special Issue, **29**.
52. J. M. Marchal and M. J. Crochet, *J. Non-Newtonian Fluid Mech.*, 1987, **26**, 77.
53. B. Debbaut, J. M. Marchal and M. J. Crochet, *J. Non-Newtonian Fluid Mech.*, 1988, **29**, 119.
54. S. Dupont and M. J. Crochet, *J. Non-Newtonian Fluid Mech.*, 1988, **29**, 81.
55. X.-L. Luo and E. Mitsoulis, *J. Rheol.*, 1990, **34**, 309.
56. B. Debbaut, M. J. Crochet, H. Barnes and K. Walters, *Proc. Xth Int. Cong. Rheol.*, 1988, **2**, 291.
57. B. Debbaut and M. J. Crochet, *J. Non-Newtonian Fluid Mech.*, 1988, **30**, 169.
58. B. Debbaut, *J. Non-Newtonian Fluid Mech.*, 1990, **37**, 281.
59. B. Debbaut, *J. Non-Newtonian Fluid Mech.*, 1990, **36**, 265.
60. K. Chiba, K. Nakamura and D. V. Boger, *J. Non-Newtonian Fluid Mech.*, 1990, **35**, 1.
61. J. H. Song and J. Y. Yoo, *J. Non-Newtonian Fluid Mech.*, 1987, **24**, 221.

62. J. R. Rosenberg and R. Keunings, *J. Non-Newtonian Fluid Mech.*, 1988, **29**, 295.
63. H. C. Choi, J. H. Song and J. Y. Yoo, *J. Non-Newtonian Fluid Mech.*, 1988, **29**, 347.
64. W. Gleissle, *Proc. Xth Int. Cong. Rheol.*, 1988, **1**, 350.
65. G. K. Batchelor, *J. Fluid Mech.*, 1971, **46**, 813.
66. J. D. Goddard, *J. Non-Newtonian Fluid Mech.*, 1976, **1**, 1.
67. W. Philipoff and F. H. Gaskins, *Trans. Soc. Rheol.*, 1958, **2**, 263.
68. D. H. Fruman and M. Barigah, *Rheol. Acta*, 1982, **21**, 556.
69. A. B. Metzner and A. P. Metzner, *Rheol. Acta*, 1970, **9**, 174.
70. A. B. Metzner, *Rheol. Acta*, 1971, **10**, 434.
71. H. P. Hurlimann and W. Knappe, *Rheol. Acta*, 1972, **11**, 292.
72. C. Balakrishnan and R. J. Gordon, *AIChE J.*, 1975, **21**, 1225.
73. A. K. Chakrabarti and A. B. Metzner, *J. Rheol.*, 1986, **30**, 29.
74. B. Tremblay, *J. Non-Newtonian Fluid Mech.*, 1989, **33**, 137.
75. F. N. Cogswell, *Polym. Eng. Sci.* 1972, **12**, 64.
76. D. M. Binding, *J. Non-Newtonian Fluid Mech.*, 1988, **27**, 173.
77. D. M. Binding and D. M. Jones, *Rheol. Acta*, 1989, **28**, 215.
78. D. M. Binding, D. M. Jones and K. Walters, *J. Non-Newtonian Fluid Mech.*, 1990, **35**, 121.
79. M. M. Denn, *Ann. Rev. Fluid Mech.*, 1990, **22**, 13.
80. E. A. Christiansen, S. J. Kelsey and T. R. Carter, *AIChE J.*, 1972, **18**, 372.
81. D. M. Binding, Paper presented at Conference on Flow Processes in Composite Materials, Limerick, Ireland, July 1991.
82. D. M. Binding, Paper presented at the 3rd Int. Workshop on Extensional Flows, Villard de Lans, France, January 1991.

Chapter 2

A Critical Appraisal of Available Methods for the Measurement of Extensional Properties of Mobile Systems

D. F. JAMES

Department of Mechanical Engineering, University of Toronto, Toronto, Canada

AND

K. WALTERS

Department of Mathematics, University of Wales, Aberystwyth, UK

2.1. INTRODUCTION

There is now a general consensus in the field of non-Newtonian fluid mechanics that the *extensional* viscosity of mobile highly elastic liquids can be a very important influence in determining flow characteristics. Depending on the complexity of the flow geometry, it can be at least as important as the more widely measured *shear* viscosity, and in some cases it may be the more dominant influence. Not surprisingly, this realization has motivated the development of extensional rheometers. The process has been reasonably successful for (stiff) highly viscous liquids like polymer melts, although it must be conceded that even in this case there can be severe limitations on the attainable range of equilibrium strain rates. The problems for mobile systems, like polymer solutions, are more acute and they are of a different nature to those encountered with stiff systems. Basically, the difficulties are associated with the need to generate flows with a well-defined extensional flow field, free of a substantial shear component.

In normal circumstances, the difficulties would be viewed as insurmountable. However, since mobile elastic liquids can often exhibit exceptionally high stress levels in extension-dominated flows, the search for viable extensional rheometers will only be abandoned with reluctance. Hence the motivation for the present chapter.

We attempt to provide a 'state-of-the-field' assessment, which will include *inter alia* a realistic and objective estimate of the value of available extensional rheometers. For more general discussions of extensional viscosity, we refer the reader to Refs 1 and 2.

2.2. BASIC DEFINITIONS

The resistance to extensional motion is usually expressed as a viscosity, analogous to the property expressing resistance to shearing motion. For extensional motion, the property is called extensional viscosity and is written as η_E, which is the most common symbol in the literature and the one in accord with nomenclature suggested by the Society of Rheology.[3]

In the same way that shear viscosity should be measured in a simple shear flow with a uniform shear rate throughout, the extensional viscosity should be measured when the deformation is purely extensional and the extensional rate is constant throughout. Purely extensional motions may be either planar or axisymmetric, but we will deal only with axisymmetric flows because these flows are more important industrially and because it is easier to make measurements of η_E in axisymmetric flows. For a homogeneous axisymmetric motion, the velocity field in Cartesian coordinates is given by

$$v_x = \dot{\varepsilon}x, \qquad v_y = -\tfrac{1}{2}\dot{\varepsilon}y, \qquad v_z = -\tfrac{1}{2}\dot{\varepsilon}z \qquad (2.1)$$

where $\dot{\varepsilon}$ is the (constant) extensional strain rate. Analogous to shear viscosity, the extensional viscosity is defined by

$$\eta_E = \frac{\sigma_{xx} - \sigma_{yy}}{\dot{\varepsilon}} = \frac{\sigma_{xx} - \sigma_{zz}}{\dot{\varepsilon}} \qquad (2.2)$$

where σ_{xx} is the normal stress in the direction of extension and σ_{xx} and σ_{yy} are the normal stresses in the two perpendicular directions. The shear viscosity involves a single tangential stress but extensional viscosity necessarily involves a *difference* of normal stresses. That is, in principle the normal stress for the numerator of η_E is the total normal stress in the x direction and this stress includes the pressure. The pressure is unknown but it can be eliminated by incorporating a second normal stress. For example, in the pulled-filament technique to measure η_E, one measures the total force or stress in a filament of fluid. But one cannot separate that part of the stress which is related to deformation from that part which is due to pressure. A force balance in the transverse direction shows that the pressure is the negative of the transverse stress (this derivation is shown in detail on pp. 38 and 39 of Ref. 4). Hence the transverse stress is introduced and the numerator for η_E is therefore a difference of stresses.

The definition of extensional viscosity given above is natural and logical, but it is very difficult to create experimental conditions such that it can be measured properly. The practical difficulty is that it is not possible to create the flow field of eqn (2.1) such that equilibrium conditions are attained. To reach equilibrium, the residence time of the fluid in a constant-$\dot{\varepsilon}$ flow field must be sufficiently long. Necessary residence times have been achieved only for some stiff liquids and have never been realized for mobile liquids, the materials of interest in this chapter. For these latter fluids, then, η_E depends upon residence time as well as on $\dot{\varepsilon}$. Moreover, in many cases $\dot{\varepsilon}$ is not constant during the residence time and thus, in general, η_E is a function of strain history.

Since steady-state values of η_E have never been achieved for mobile systems, any values of η_E given in the literature must be recognized as *transient*. If η_E depends upon strain history, then the preferred history is the one for which the fluid is initially undeformed and then is subjected to a constant extensional rate $\dot{\varepsilon}$. That is, the imposed strain rate should have the form

$$\dot{\varepsilon}H(t) \qquad (2.3)$$

where $H(t)$ is the Heaviside step function. The resulting 'viscosity' must then be a function of time t as well as strain rate $\dot{\varepsilon}$. Hence, for this strain history, it is proper to write

$$\eta_E = \eta_E(\dot{\varepsilon}, t) \tag{2.4}$$

The steady-state extensional viscosity is then the limiting value of η_E as $t \to \infty$, provided, of course, that an equilibrium value actually exists. It was suggested earlier[2] that this special value might also be termed the 'asymptotic extensional viscosity'.

In practice it is rarely possible to generate an extensional flow field instantaneously from a state of rest, and some prior deformation (usually called 'prehistory') must be accepted. Furthermore, generating a *constant* strain rate is far from easy, even for a short period of time. Accordingly, in most techniques the effects of both prehistory and variable strain rate have to be incorporated. In the light of these considerations, it is appropriate to ask a number of searching questions before any published extensional viscosity data are taken seriously in a comparative exercise, as follows:

  (i) Does the flow field approximate to that given in eqn (2.1) and, if not, is the variability in $\dot{\varepsilon}$ taken into account?
 (ii) If the strain rate is not constant for $t > 0$, is some averaging procedure for the calculation of η_E meaningful? (It transpires that such a procedure is possible and must often be done, but meaningfulness is another matter.)
(iii) Whether or not the strain rate is constant for $t > 0$, does the experiment or technique allow sufficient residence time for steady-state (equilibrium) stresses to be reached? If not, is the residence time known and included as a variable in the presentation of results?
 (iv) If the flow is not given by eqn (2.3), what is the effect of prehistory? That is, what are the stresses for $t \leq 0$ and how do they affect the stress response for $t > 0$?

The above questions have already anticipated some of the problems which arise in the techniques to determine extensional viscosity. In the remainder of this chapter, it will become evident that these concerns are easily warranted.

2.3. THE PROBLEM STATED: DISPARITY OF RESULTS FROM DIFFERENT EXTENSIONAL RHEOMETERS

The growing realization of the importance of extensional viscosity and the increase in techniques to measure this property were the motivation for a series of International Workshops on the subject during the period 1988–91. These workshops evolved from an earlier series of meetings on extensional flow of viscoelastic liquids. From the initial series it became apparent that the most pressing problem in extensional flow was a proper measurement of the relevant fluid property. Accordingly, the subsequent meetings focused on extensional rheometers and the techniques to measure extensional viscosity. The first two Workshops with this focus were held at Chamonix in 1988 and at Combloux in 1989, and the papers from these meetings were published as special issues of the *Journal of Non-Newtonian Fluid Mechanics*, Volumes 30 and 35, respectively.[5,6] The third Workshop was held in Villard de Lans in January 1991, but no proceedings are expected from that meeting.

During the Chamonix meeting, it was evident that the techniques to measure η_E could not be compared or evaluated because the test fluids were different, viz., each experimenter presented η_E data for a fluid of interest to him and him alone. It was realized that a common test fluid was necessary and Dr T. Sridhar, of the Chemical Engineering Department of Monash University, offered to make a fluid and to supply it to interested research groups around the world.

The chosen test fluid, designated M1 (after Monash), was a 0·244% solution of polyisobutylene in a mixed solvent consisting of 93% polybutene and 7% kerosene. It was a Boger-type fluid since it was highly elastic and had a nearly constant shear viscosity over a wide range of shear rates. Its viscosity was moderately high, being about 3 Pa s at 20°C.

Measurements on extensional viscosity of M1 were reported a year later at the next Workshop at Combloux and were published in the special issue of the *Journal of Non-Newtonian Fluid Mechanics*.[6] In that issue research groups generally reported their results in the form of η_E *versus* $\dot{\varepsilon}$, although some showed how η_E depended upon other kinematical variables. The η_E data from the special issue are presented together in Fig. 2.1. Each curve on the graph is a best-fit curve through data from the identified group. Individual references are not given for each set of data because all appear in the special issue.[6] This plot, which is effectively a summary of the Workshop and of the special issue, has not appeared before.

In view of the disparity of results in Fig. 2.1, a number of issues arise and we should first address the issue of accuracy. Accuracy varies from technique to technique, but it is possible to make a blanket estimate without knowing the details of individual techniques. Measurements of η_E and $\dot{\varepsilon}$ are likely to be accurate to no worse than ±30% because $\dot{\varepsilon}$ can usually be measured to ±20% and because $\sigma_{xx} - \sigma_{yy}$, the numerator for

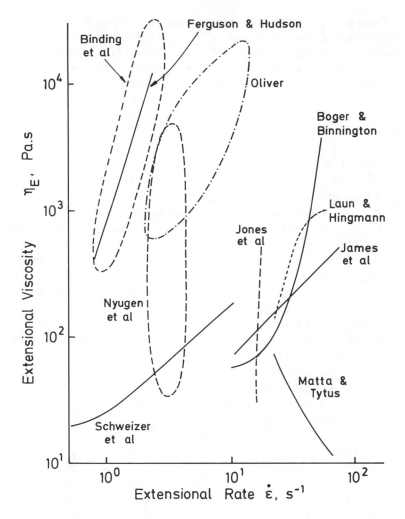

Fig. 2.1. Extensional viscosity measurements on the M1 fluid. The lines and envelopes represent data by the various M1 investigators, as published in the special M1 issue of the *Journal of Non-Newtonian Fluid Mechanics.*[6]

η_E, can usually be measured to $\pm 10\%$. These error estimates are thought to be generous (and therefore unfair to some experimenters) but better estimates are not necessary because they cannot explain the three-decade variation in Fig. 2.1.

What Fig. 2.1 actually reveals is that η_E is not a function of $\dot{\varepsilon}$ alone and therefore the reported values of η_E are not steady-state values. Research groups presented their data as η_E *versus* $\dot{\varepsilon}$ perhaps because they vaguely hoped that flow conditions might approach those of steady state and therefore that η_E might depend only weakly upon factors other than $\dot{\varepsilon}$. But the master plot clearly shows that the other factors cannot be ignored. In fact, one would be hard-pressed to make even order-of-magnitude estimates for steady-state η_E values, at any extensional rate, based on the results in the figure. Figure 2.1 does tell us, however, that M1 is a strongly elastic liquid because the η_E values are well in excess of 9 Pa s or three times the shear viscosity, the value for an inelastic fluid.

The values of η_E in Fig. 2.1 are transient values and the disparity makes it clear how strongly these values depend upon variables other than strain rate. Before the workshop, it was recognized that transient values of η_E depend upon strain and prestraining, as well as on strain rate, but few expected that the dependence upon the other variables would be so strong as to produce the differences in Fig. 2.1. As will be seen later, none of the techniques creates the desired flow field $\dot{\varepsilon}H(t)$, and some have significant straining prior to the purely extensional motion. Since Fig. 2.1 does not take into account either pre-straining or strain history during the extensional motion, this figure is really a 'snap-shot' of η_E values: the values are captured at particular levels of $\dot{\varepsilon}$ but these occur at different times during different strain histories.

The η_E data from the different instruments can properly be compared only for the same strain histories or when the differences in strain histories can be accounted for. As subsequent sections will show, no two instruments have the same history, although a new one promises to provide a history close to $\dot{\varepsilon}H(t)$. No one has so far tried to account for prehistory, but Keiller[7] and Mitsoulis[8] have analyzed the M1 data for different extensional histories. Since these analyses are best understood once the measurement techniques are explained, a description of one of these works (Keiller's) is included in Section 2.5.

After the Combloux meeting, a similar international exercise was undertaken with another test fluid, this time a shear-thinning fluid. It was a solution of polyisobutylene in decalin, prepared by Prof. K. Walters of Aberystwyth and thus appropriately designated A1. The techniques used to measure η_E for A1 were essentially the same as those for M1, and the results were reported at the Villard de Lans Workshop in 1991. The $\eta_E(\dot{\varepsilon})$ data for A1 were also disparate, but somewhat less so this time, perhaps because A1 was not as elastic as M1.

Leaving aside for the moment the thorny question of which of the available rheometers provides the nearest estimate of the true (equilibrium) extensional viscosity, we can at least make some positive comments on the current situation.

(i) The majority, if not all, of the rheometers should be viewed as providing measurements of *an* extensional viscosity rather than *the* extensional viscosity.

(ii) The rheometers may nevertheless be extremely useful for quality control in industry, for they can provide a convenient ordering of fluids with respect to their resistance in extension-dominated flows.

(iii) It is essential to specify both the prehistory before the extensional-flow test section and the duration of the extensional deformation itself and to take these factors into account when interpreting data from the different instruments.

(iv) Some of the existing techniques can play a useful role in developing constitutive equations for test fluids, particularly those techniques with well-defined flow fields. Because the two test fluids (M1 and A1) were well characterized in shear, data for those fluids in well-defined stress and strain fields are especially useful in identifying the constitutive equations most suitable for the test fields. It is anticipated that data from these flows will play an increasingly important role in 'process modelling'.

Having set the scene, as it were, it is now appropriate to consider the various types of extensional rheometers which are currently available. In the light of the foregoing discussion, it is clearly important to view the associated kinematics in a critical fashion. It will not have escaped the attention of the perceptive reader that the two authors of this chapter have invested much time and effort in developing a number of the techniques discussed below. We shall (with some difficulty) attempt to be objective.

2.4. EXTENSIONAL RHEOMETERS

The laboratory techniques which have been devised to measure extensional viscosity are described below. The commercial instruments which have been developed from two of these will be discussed separately. A recent article by Gupta and Sridhar[9] reviews the techniques that were available up to 1988 for both stiff and mobile systems. Their review is very useful in delineating the characteristics of measured flow variables: attainable ranges, recording techniques, accuracy, sensitivity and stability. The present chapter will adopt a different, but complementary viewpoint. Of particular concern are the straining histories and how closely they approximate to the ideal history. This concern will lead to investigations of the shearing levels prior to and during the extensional motion and to an assessment of their effects on the measured values of η_E.

2.4.1. Spinning Devices

The most-developed method of measuring η_E for mobile liquids is akin to the spinning process for plastic fibres. It consists in drawing out a fluid filament, as shown in Fig. 2.2. The fluid is drawn down from an orifice or tube by a rotating drum (if the spinning forces are high enough), as in Ferguson's original design,[10] or by a suction device of the sort pioneered by Gupta and Sridhar.[11]

A steady downward flow can usually be generated for a restricted range of conditions. The kinematics are not difficult to determine from the measured flow rate and the dimensions of the fluid filament. The latter can be determined with reasonable accuracy using a video-camera technique or the equivalent. The stress is readily found from the filament area and the tension within it, the latter determined from a measurement of either the torque in the drum or the force on the exit tube. Accurate determinations of the kinematics and the stress field are not difficult in this technique. Problems lie elsewhere.

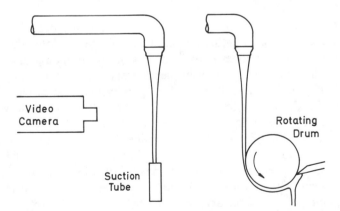

Fig. 2.2. Schematic diagram of the spinning technique. The fluid issuing from the nozzle is pulled downward by a suction tube or rotating drum.

Firstly, varying the strain rate in the thread is not easy and hence the extensional strain-rate window with this technique can be rather narrow. This limitation arises because the threadline geometry is not constant but is related to the flow rate and to other instrument variables. For example, increasing the flow rate (other things being equal) generally results in a thicker overall thread and the increase in strain rate is disappointingly smaller than the imposed increase in flow rate. The limited variation in strain rate is borne out by the steep lines for data from this technique in Fig. 2.1.

However, the *main* disadvantages of the fibre-spinning methods are related to the strain history. Under special circumstances, which cannot be deduced or manufactured *a priori*, a relatively constant extensional strain rate exists along the filament. Consequently, in most cases the flow is manifestly unsteady *in a Lagrangian sense* and far removed from the idealized flow envisaged in eqn (2.1). Furthermore, even when a constant strain rate does exist along a major portion of the threadline, prehistory in the upstream tube and orifice and in any die-swell region immediately after the orifice can have a marked effect on the stress field.

A further problem with the flow field needs to be mentioned. It concerns the possibility of a significant and largely unexpected shearing component in the filament. Using small particles to monitor local velocities, Ferguson and Missaghi[12] detected a transverse variation in axial velocity and observed that the slower-moving particles were located on the periphery of the filament. Consequently, what appears to be a relatively simple technique is actually quite complicated from a fluid-mechanics point of view.

Determining values of η_E and $\dot\varepsilon$ from spinline data has been undertaken at specific points along the threadline, or else made in a global (averaged) sense, as favoured by Jones *et al.*[13] In view of the foregoing discussions, nothing is lost by adopting the simpler averaging procedure. In this, the complicating effects of surface tension, gravity and inertia can also be accommodated, at least to a sufficient degree within the acknowledged limitations of the technique.

It should also be pointed out that spinline rheometers of the same type yield similar data. In Fig. 2.1, the data of Binding *et al.* and of Hudson & Ferguson were obtained with the same type of spinline rheometers and the agreement of the data sets is good. This agreement is consistent with our previous assertion (in Section 2.3) that the accuracy of data from extensional rheometers is no worse than about ±30% for a particular instrument.

In summary, the problems with spinning devices are:

(i) the limited range of extensional strain rate;
(ii) the non-ideality of the flow field, viz. the extensional strain rate is generally not constant along the filament and significant shear can be present;
(iii) the major effect of preshear in the upstream tube and orifice flows.

Notwithstanding these difficulties, spinning devices have been (and will continue to be) useful in quality control and in providing an idea of extensional-viscosity levels for mobile (non-concentrated) solutions of flexible high-molecular-weight polymers. These levels are easily measurable with these simple devices because the levels can sometimes be several orders of magnitude higher than those expected for comparable Newtonian liquids.

A different but related spinning technique is Matta's falling-bob apparatus,[14] as illustrated in Fig. 2.3. The device is a noteworthy attempt to overcome some of the difficulties in the spinline device.

The technique consists of placing a drop of fluid between the ends of two cylindrical bobs, one of which is stationary; the other is allowed to fall under gravity to pull the drop out and into a thread. Photographs of the fluid show that, at any instant of time during the fall, the fluid is a filament of constant diameter except for the small attachment zones at the ends. Hence almost all of the fluid sample experiences the same purely

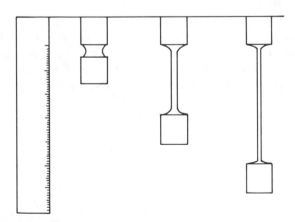

Fig. 2.3. Matta's falling-bob technique for measuring extensional viscosity. A purely extensional motion is created in the fluid filament.

extensional motion. The major drawback of the technique is that the
strain-rate history is uncontrollable and depends upon the material under
test.

Because the fall of the bob depends upon the tension in the thread, the
bob acceleration decreases with time and thus the extensional rate does
likewise. But the stress in the filament increases with time, as the material
is stretched out. Hence the extensional viscosity appears to decrease with
extensional rate, as Fig. 2.1 shows. These data, more than any other,
demonstrate the inappropriateness of plotting η_E *versus* $\dot{\varepsilon}$ when η_E
depends upon strain (or duration) as well as strain rate.

An improved device of this type has recently been designed and built
by Sridhar *et al.*[15] In this device, the descent of the bob is controlled so
that the rate of extension is constant, a technique similar to the spinning
experiment of Fernandez-Luque.[16] With a programmed fall in the Sridhar
device, it appears that the strain-rate history is close to the ideal expressed
in eqn (2.3). However, it should be pointed out that the fluid does not
change from a squashed droplet to a cylindrical filament instantaneously.
The transition from rest to uniform $\dot{\varepsilon}$ requires one unit, or more likely
several, units of strain. This amount of strain during the ramp-up is not
insignificant, but must be considered acceptable in light of the fact that
there is no prehistory and that the motion is closer to the ideal than in
any other device.

2.4.2. The Open-Syphon Technique (Fano Flow)

Fano flow or the tubeless syphon flow is shown schematically in Fig. 2.4.

The operation of a normal syphon (for simple fluids) is well known,
but for mobile, highly elastic fluids the end of the syphon can be located
several centimetres above the level of the liquid without impeding the
syphon action. It is apparent that Fano flow is similar to the spinline
flows and the relevant flow variables can be determined in much the same
way.

One major advantage is that the fluid is initially undeformed. However,
like the spinline flows, there is significant shearing in the flow. There is a
stress-free boundary everywhere but, conversely, it is not shear-free except
perhaps in the nearly-cylindrical upper portion. Shear is generated
because the streamlines are curved and the boundary has zero stress (as
explained in Ref. 2). The flow visualization work of Matthys[17] confirms
the presence of shear in the curved part of the flow.

In conclusion, Fano flow suffers from similar disadvantages to the
spinning devices, except that it has a more predictable and acceptable

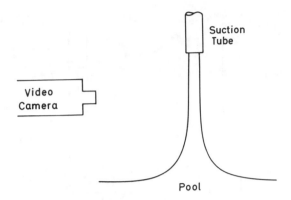

Fig. 2.4. Schematic diagram of Fano flow.

preshear history. Like other techniques for measuring extensional flow resistance, it is certainly a useful technique for quality control purposes (although the relevant strain rate window is again very narrow) and in assessing constitutive equations.

2.4.3. Stagnation-Point Flow Techniques

An interesting set of methods may be classified under the generic heading 'stagnation-point flow techniques'. These have been pioneered by Keller and Odell,[18] Laun and Hingmann,[19] Fuller et al.[20] and others. As illustrated in Fig. 2.5, fluid is drawn into (or issues from) opposing tubes to create a stagnation flow about the midplane.

One inlet is stationary and the other is part of a balance arm to measure the force exerted by the entering liquid. Such an instrument has some advantages. It is relatively simple to operate and there is no strong restriction on the viscosity levels which can be handled. Furthermore, the test liquids do not need to be 'spinnable'.

A drawback of the technique is that the strain history is not the same for all particles, even though the extensional rate in the stagnation zone is constant. The closer a particle is to the stagnation point, the higher is its residence time in the flow.

Another drawback is related to the stability of the presumed flow. A definite bifurcation in the flow was observed for the M1 fluid, at relatively low extension rates, resulting in a significant reduction in the stress levels compared with those which would have resulted from the original flow

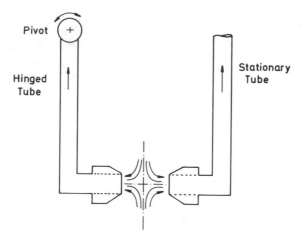

Fig. 2.5. The stagnation flow created by opposed entry tubes (see, for example, Ref. 20).

pattern.[19,20] But, even before the bifurcation occurs, the flow pattern probably changes because non-Newtonian normal stresses are induced in the fluid. Hence, the strain history is probably qualitatively, as well as quantitatively, different at different strain rates.

A related concern about the flow field arises from an analysis carried out by Schunk and Scriven.[21] Using finite element analysis, they determined the velocity and strain-rate fields around the opposed entry tubes for a Newtonian fluid. From the latter field they found relative rates of rotation throughout the flow field and thus were able to identify the proportions of shear and extension everywhere. Close to the stagnation point, the flow is purely extensional, as expected, but the region is smaller than one might expect. In fact, the fluid enters the sharp-edged entry tube in a state which consists much more of shear than of extension. Hence, overall, the fluid is not subjected to purely extensional motion, despite the stagnation point. It is not yet clear if this situation is improved when the fluid is shear-thinning or elastic. Hence, some work is required to clarify the strain history in stagnation point flows.

This technique, which promised much, must now be placed in the same category as those in Sections 2.4.1 and 2.4.2, with quality control and constitutive-equation determination being again the most likely applications of the technique in the long term.

2.4.4. Contraction-Flow Devices

There is no doubt that, from the standpoint of ease of experimentation, sudden-contraction flow devices hold significant advantages. The basic technique is simple and is shown schematically in Fig. 2.6. In many cases, especially when the test fluid is highly elastic, so-called vortex enhancement occurs on the upstream side of the contraction or orifice, and the flow in the central core is primarily extensional. A knowledge of the precise flow pattern is not required in some interpretations of the data but, when the pattern is known, it and the measured mean flow rate yield a simple measure of global extensional strain rate. The relevant dynamic measurement is provided by a pressure transducer which measures the pressure drop across the orifice plate or sudden contraction. In the latter case, the upstream tap to the transducer should be located before the vortex, i.e. before the flow starts to converge.

The technique was pioneered by Cogswell[22] and has been refined and improved by Binding [Refs 23, 24 and Chapter 1 in this book]. One of its main attractions is that the required measurements for extensional viscosity determination are already known through the so-called 'Bagley

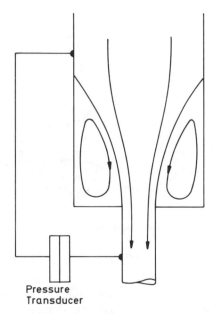

Pressure
Transducer

Fig. 2.6. Schematic diagram of contraction flow.

correction', which is applied when capillary rheometers are used to yield the *shear* viscosity. This end correction will clearly be of importance to those rheologists who routinely employ capillary rheometers to measure shear viscosity.

Because the converging contraction flow is complex, there is considerable interest in the assumptions required for data interpretation. Validating the assumptions will probably require the application of sophisticated numerical techniques for complete resolution and the present authors will not be surprised if the indefinite article 'an' will again be applied to 'extensional viscosity' for this technique.

2.4.5. Converging-Channel Flow

An instrument based on converging-channel flow has recently been pioneered by James *et al.*[25]† Its primary component is a converging channel of a shape such that the extensional rate is constant in the central region. An important aspect is that the flow is at high Reynolds numbers so that shear effects are confined to the wall region. The rheometer is shown schematically in Fig. 2.7.

For the measurement of η_E for elastic liquids, one requires knowledge of the behaviour of the *inelastic* component of the fluid and this must be

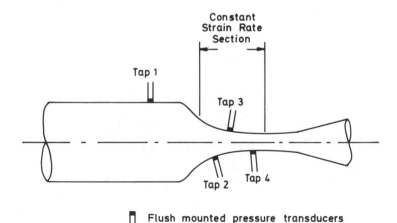

Fig. 2.7. Schematic diagram of the James converging-channel technique.[25]

† This technique should not be confused with the *lubricated* converging-flow rheometer developed by Walters and co-workers.[1,13] The amount of test fluid required (about 100 litres) and the difficulty of operation have rendered the technique obsolete. Other lubricated techniques are also no longer in vogue.[1]

provided by theoretical means using the known shear viscosity behaviour of the test fluid and an assumed flow field.

The following points about this technique are in order. Firstly, a knowledge of the second normal stress difference N_2 is required in the interpretation of the experimental data and this quantity is very difficult to measure on a routine basis. We remark that accurate measurements of N_2 are available for the test fluids M1 and A1, so there is no problem for these important fluids. However, there is no standard instrument for routine measurements of N_2 and in general it is necessary to assume some functional behaviour for N_2. The relationship $N_2 = -0.1N_1$, where N_1 is the first normal stress difference, would seem to be an acceptable means of estimating N_2 for many polymeric systems,[1] particularly because N_1 can generally be measured. Since the term involving N_2 in the data interpretation contributes from 10% to 50% of the relevant normal stress difference, accuracy for η_E clearly depends upon measured or estimated values of N_2. However, since one significant figure still represents an achievement, a general lack of N_2 data is not crippling.

Because the apparatus operates at Reynolds numbers of order 10^2, it is therefore must suitable for low-viscosity fluids and for data at high strain rates. A boundary-layer concept is introduced for this analysis and a major assumption (which has yet to be verified) is that the boundary-layer characteristics are not unduly modified by viscoelasticity. If this assumption is vindicated, flow in the converging-channel rheometer will provide an approximation to the Heaviside strain-rate assumption (eqn (2.3)) and the technique will then be useful to those who are interested in the (esoterically) important transient extensional viscosity function $\bar{\eta}_E$ $(\dot{\varepsilon}, t)$. Since the instrument is not simple, it is less attractive than other devices for quality-control purposes.

2.4.6. Other Techniques

A glance at the recent literature[5,6] would soon identify other novel techniques for extensional-viscosity measurement. However, on the whole, the kinematics associated with these techniques are not well defined and the devices are sometimes difficult to operate. These additional techniques must therefore be viewed as supportive to the popular techniques discussed above, with little or no hope of being adopted in routine studies.

2.4.7. Commercial Rheometers

At the time of writing, there are two commercial extensional rheometers on the market and each was developed from a technique described earlier.

In each case, the technique originated in a university laboratory and was developed into a commercial instrument.

The Carri-Med Elongational Viscometer, sold by the Carri-Med Corp. of Dorking, UK, is based on Ferguson's spinline technique.[10] This instrument, which is sometimes referred to as a spinline rheometer (SLR), costs about £33 000 and is useful for room-temperature fluids which have viscosities in the range 0·1–100 Pa s. The device is straightforward to operate and, with reasonable attention paid to variations in ambient temperature, provides reproducible data.

A video system for measuring filament diameter is now available from the manufacturer. This equipment is indispensable because the filament diameter is small (about 1 mm) and changes slowly, and the differences in diameter must be precisely known to determine the rate of extension. Hence accuracy in determining $\dot{\varepsilon}$ is difficult. Another limitation of the instrument, as mentioned in Section 4.1, is that η_E can be measured only over narrow ranges of the extensional rate.

The other extensional rheometer is Rheometrics (Piscataway, New Jersey, USA) 'Fluid Analyzer RFX', costing about £30 000. Introduced at the beginning of 1991, the device is based on Fuller's opposed-entry-tube design (Fig. 2.5). Rheometrics has extended Fuller's original design so that the instrument can measure shear as well as extensional flow resistance. For the shear mode, the entry tubes are replaced by small-bore tubes, and therefore flow resistance is measured at high shear rates. It must be pointed out that this mode is not a substitute for a proper measurement of shear viscosity because of entrance effects in the tubes. Even for Newtonian fluids, a sufficient entry length is required to develop fully the flow field. For viscoelastic fluids, an additional entrance effect is that flow into the tube creates an extensional flow field and therefore an extra flow resistance. Hence the instrument measures 'apparent shear viscosities', as stated in the Rheometrics literature. Similarly, when the opposed entry tubes are attached for extensional flow measurement, the device measures 'apparent extensional viscosity'. In comparison to the SLR, the RFX can measure η_E over a wide range of extensional rates, the greater range being evident from the curves in Fig. 2.1. Furthermore, since test materials do not have to be spinnable for the RFX, it can be used for less viscous fluids.

The two extensional rheometers provide very different values for η_E for the same fluid, as Fig. 2.1 demonstrates. Neither instrument yields a true η_E, as noted earlier, but both yield assessments of fluid behaviour which are useful for quality control. If an absolute value of η_E is needed for a particular purpose, it is difficult to choose between the values provided by the instruments. Perhaps the best advice is to choose the value from the instrument whose flow more closely matches the flow in the process of interest. If, for example, the latter contains a great deal of shearing prior to extension, then the value from the SLR is likely to be more reliable.

However, for absolute measurements of η_E, we await the development of new instruments.

2.5. CONSTITUTIVE MODELLING OF THE M1 TEST FLUID

Because of the difficulty in making meaningful measurements in exten sional flow—Fig. 2.1 being a spectacular demonstration of the difficulty— it is not surprising that there has been little success in establishing which constitutive equations, if any, are most appropriate for modelling extensional flow fields. This situation is in contrast to that for shearing flows where, for various fluids and for widely varying shear flows, there has been considerable success in matching constitutive equations to experimental data. For the fluid M1, for example, the Oldroyd model fits the viscometric data because the viscosity is nearly independent of shear rate and because the first normal stress difference in shear is quadratic in shear rate.[6]

Because the Oldroyd model fits the shear properties of M1, it was natural to start with this model when investigating models which might be suitable for the various extensional flow fields. When Fig. 2.1 was presented earlier, it was mentioned that Keiller[7] analyzed the M1 data and found some consistency based on the Oldroyd model. Keiller shows that the model can simulate the results from several M1 experiments, quantitatively as well as qualitatively. In particular, the model predicts:

(i) the kinematics of the falling drop experiment of Jones *et al.*,[26] particularly the nature of the fluctuations of strain rate with time;

(ii) the fall distance as a function of time for Matta's falling-bob technique;

(iii) the dependence of stress on strain and travel time for Oliver's pulled filaments.

Further evidence that the Oldroyd model may be the most appropriate one for M1 is provided by Nguyen *et al.*[27] and by Chai and Yeow.[28] Both groups were able to fit spinline data for M1 to predictions from the

Oldroyd model when initial stresses at the top of the spinline were appropriately chosen.

It is important to point out that Chai and Yeow also proposed the use of a three-relaxation-time KBKZ model to represent the viscometric and dynamic data for M1, the agreement with experimental data being an improvement over the Oldroyd B model. Interestingly, Mitsoulis[8] recently employed the KBKZ model in a numerical simulation exercise for abrupt-contraction flow, with encouraging results.

The above developments may be viewed as potentially important, being indications that existing extensional-flow techniques may be used to construct continuum models for test fluids.

2.6. CONCLUSIONS

This chapter has attempted to highlight the current situation of instrument diversity and also the pressing requirements which will facilitate an objective assessment of instrument utility. These requirements may be expressed thus:

(i) For the sudden-contraction devices, a numerical experiment is urgently needed to estimate the effect of shear on the interpretation of the data and also to examine the underlying assumptions in the data interpretation.

(ii) For the converging-flow technique, viscoelastic effects in the boundary layer need to be clarified, either by velocity measurements or by numerical analyses of the stress and velocity fields.

(iii) For the opposed-tube technique, more work on the effect of changing instrument dimensions would be in order, together with a careful study of induced stresses and their effect on flow pattern, flow stability and data interpretation.

The above list is not meant to be exclusive, and there is certainly scope for more theoretical and experimental work on the general subject. However, after three comprehensive workshops on the subject, the field is now at a crossroad. A number of techniques will survive and will prove to have utility both in academia and industry, provided that the claims for each are limited. We shall undoubtedly hear more rheologists talking of *an* extensional viscosity arising from their work, rather than *the* extensional viscosity, and this is to be applauded.

There is no doubt that some of the available viscometers will become increasingly important in quality control, and some will find use in constitutive equation development and process modelling.

REFERENCES

1. H. A. Barnes, J. F. Hutton and K. Walters, *An Introduction to Rheology*, Elsevier, Amsterdam, 1989, Ch. 5.
2. D. F. James, in *Recent Developments in Structured Continua, Vol 2*, D. De Kee and P. N. Kaloni (Eds), Longman, Essex, 1990, Ch. 12.
3. J. M. Dealy, *J. Rheology*, 1984, **28**, 181.
4. R. B. Bird, R. C. Armstrong and O. Hassager, *Dynamics of Polymeric Liquids, Vol 1*, Wiley-Interscience, New York, 1987.
5. *J. Non-Newtonian Fluid Mech.*, 1988, **30**, 97.
6. *J. Non-Newtonian Fluid Mech.*, 1990, **35**, 85.
7. R. Keiller, *J. Non-Newtonian Fluid Mech.*, 1992, **42**, 49.
8. E. Mitsoulis, *J. Non-Newtonian Fluid Mech.*, 1992, **42**, 30.
9. R. K. Gupta and T. Sridhar, in *Rheological Measurement*, A. A. Collyer and D. W. Clegg (Eds), Elsevier, Applied Science, London, 1988, Ch. 8.
10. N. Hudson and J. Ferguson, *Trans. Soc.*, 1976, **20**, 265.
11. R. K. Gupta and T. Sridhar, *Advances in Rheology. Vol 4. Applications*, B. Mena *et al.* (Eds), Univ. Nacional Autonom de Mexico, Mexico, 1984, 71.
12. J. Ferguson and K. Missaghi, *J. Non-Newtonian Fluid Mech.*, 1982, **11**, 269.
13. D. M. Jones, K. Walters and P. R. Williams, *Rheol. Acta*, 1987, **26**, 20.
14. J. Matta and R. P. Tytus, *J. Non-Newtonian Fluid Mech.*, 1990, **35**, 215.
15. T. Sridhar, V. Tirtaatmadja, D. A. Nguyen and R. K. Gupta, *J. Non-Newtonian Fluid Mech.*, 1991, **40**, 27.
16. R. Fernandez-Luque, to be published.
17. E. F. Matthys, *J. Rheology*, 1988, **32**, 773.
18. A. Keller and J. A. Odell, *Colloid Polym. Sci.*, 1985, **263**, 181.
19. H. M. Laun and R. Hingmann, *J. Non-Newtonian Fluid Mech.*, 1990, **35**, 137.
20. G. G. Fuller, C. A. Cathey, B. Hubbard and B. E. Sebrowski, *J. Rheology*, 1987, **31**, 235.
21. P. R. Schunk and L. E. Scriven, *J. Rheology*, 1990, **34**, 1085.
22. F. N. Cogswell, *Polym. Eng. Sci.*, 1972, **12**, 64.
23. D. M. Binding, *J. Non-Newtonian Fluid Mech.*, 1988, **27**, 173.
24. D. M. Binding and K. Walters, *J. Non-Newtonian Fluid Mech.*, 1988, **30**, 233.
25. D. F. James, G. M. Chandler and S. J. Armour, *J. Non-Newtonian Fluid Mech.*, 1990, **35**, 421.
26. W. M. Jones, N. E. Hudson and J. Ferguson, *J. Non-Newtonian Fluid Mech.*, 1990, **35**, 263.
27. D. A. Nguyen, R. K. Gupta and T. Sridhar, *J. Non-Newtonian Fluid Mech.*, 1990, **35**, 207.
28. M. S. Chai and Y. L. Yeow, *J. Non-Newtonian Fluid Mech.*, 1990, **35**, 459.

Chapter 3

Surface Rheology

B. WARBURTON

The School of Pharmacy, University of London, London, UK

3.1. INTRODUCTION AND HISTORY

In addition to the well-known and traditional bulk rheology, there exists the possibility of measuring rheological properties at the surface of a solution or at the liquid/liquid interface (say the oil/water interface).

This surface rheology is not solely of academic or theoretical interest. In the 1920s and 1930s, it was largely a specialized and fundamental pursuit. In the last ten years, the applications or potential applications in industry of this technique, as another weapon in the armoury of the physical scientist, are simply enormous.

Originally, there was a lot of discussion as to the true nature of a surface rheological investigation. The reason is that the surface rheological parameters for pure liquids are very low indeed compared to the values obtained in the presence of an absorbed or deposited surface film of another component. The other problem, as will be seen, is the fact that for many surface rheological techniques, it proves almost impossible to separate a contribution to the supposed measurements from the substrate or bulk.

The presence of thin films of second components, e.g. surfactants, long-chain aliphatic alcohols or polymers, can be considered in molecular terms as being in the form of either a monolayer or multilayer.

Fundamentalists considered only the molecular monolayer to be strictly relevant to the subject of surface rheology. Historically the monolayer had been considered thermodynamically by Gibbs as a separate phase and any formal treatment of this would need precise definitions. However, modern technological applications of thin films in electronics, biology, medicine and pharmaceuticals demand knowledge of the rate of formation and mechanical properties of molecular multilayers of second and even third components. If one is careful with definitions and underlying assumptions, there need be no loss of rigour when surface rheological methods are applied to molecular multilayers, as a thickness parameter and a homogeneity parameter[1] can be used to give the correct dimensionality of two to the measurements.

The more serious problem is the concurrent existence of the bulk properties of the underlying medium and it is not always either theoretically or practically possible to separate these bulk properties from the surface properties.

As with bulk rheology, we have the opportunity of performing experiments either in elongation or in compression, i.e. *dilatation* and shear. In general terms, we would expect to find all those particularly characteristic rheological behaviours as Newtonian flow, Hookean elasticity and linear viscoelasticity in the surface as in the bulk. The more complex behaviours of plastic yield, thixotropy and rheopexy are also found. However, perhaps unexpectedly, surface rheological properties can exhibit pronounced ageing properties *anisotropechrony*[1] which must be taken on board in any plausible theoretical model.

A few years ago, rheologists were divided into two camps: those looking at systems as essentially structureless continua and those looking at systems as ensembles of molecules.

However, the work of Byron Bird[2] and others at Wisconsin has allowed the possibility of 'having the cake and eating it' in this respect. The molecular problems become really most acute when dealing with polymer solutes. As Bird and his associates point out in their preface: 'Understanding polymer fluid dynamics is important in the plastics industry, performance of lubricants, application of paints, processing of foodstuffs and movement of biological fluids'. The idealized molecular models of polymers consist mainly of rigid dumbbells for rodlike polymers and bead-and-spring models for more flexible polymers. Most of the published work in this area has been addressed to the bulk phase, and clearly the rheology of the surface is as yet an unexplored chasm. The foundations of this projected work must surely reflect the physical chemistry of surface kinetics and surface thermodynamics. Silberberg's classical contributions[3-5] are still authoritative for polymer and polyelectrolyte adsorption.

As mentioned earlier, surface rheology has considerable potential both in biology and medicine. In 1945, J. F. Danielli[6] wrote a paper to *Nature* entitled 'Reactions at interfaces and their significance in Biology'. He suggested that before biologists could consider the theoretical possibilities of such work two requirements must be met: (1) techniques should be devised capable of dealing with the surface properties of biological molecules and (2) an explanation should be made of the types of reaction and behaviour characteristics of molecules at interfaces. We are now fully aware of the relevance of surface rheology to biological processes. Examples of these are immunological and enzymatical[7] and lubricational.[8,9]

3.2. DEFINITIONS AND THEORY

We shall now proceed with the definitions of surface stress, strain and rate of strain, in both elongation (dilatation) and shear. Although we shall be discussing both the surface and the interface, for the sake of brevity, only the term surface will be used to cover the liquid 1/air interface and the liquid 1/liquid 2 interface. The axes will be labelled as a right-hand configuration 1, 2 and 3, where 1 and 2 are in the horizontal surface plane and 3 is perpendicular or transverse. These directions are sometimes referred to as x, y and z or even $\mathbf{i}$, $\mathbf{j}$ and $\mathbf{k}$ if the writer is using vectors, where the vector set $\{\mathbf{ijk}\}$ is of arbitrary direction in 3-space.[10]

3.2.1. Surface Stress
The total surface stress tensor's physical components will number four and there will also be a functional dependence on the surface tension. There will be two *normal* and two *shear* components.

According to Vermaak et al.:[11]

$$p_{ij} = \delta_{ij}\gamma + \frac{\partial\gamma}{\partial\varepsilon_{ij}} \tag{3.1}$$

where

$$\delta_{ij} = \text{kronecker Delta}$$

$$= 1 \quad \text{when } i=j$$

$$= 0 \quad \text{when } i \neq j$$

$$\gamma = \text{surface tension}$$

$$\varepsilon_{ij} = \text{surface strain tensor}$$

$$p_{ij} = \text{surface stress tensor}$$

However, an extra term will be required if a solid elastic field of insoluble component is also present at the surface.

For dilatation in the 1 direction:

$$p_{11} = \gamma + \frac{d\gamma}{d\varepsilon_{11}} + E_{ps}\varepsilon_{11} \tag{3.2}$$

where E_{ps} is the usual surface Young's Modulus of film material and not the Gibbs elasticity. In the shear case:

$$p_{12} = G_{ps}2\varepsilon_{12} \, \dagger \tag{3.3}$$

The significance of the coefficient 2 in eqn (3.3) arises in a similar way to the treatment of shear in the bulk, and the reader is referred to Sherman,[12] for a full explanation.

It is important to grasp the role played by surface tension in these considerations. The concept of surface tension *in two dimensions* is close to the concept of pressure *in three dimensions*. The pressure in, say, a steam boiler, will have the same units $mN\,m^{-2}$ as the stress tensor's physical components, but pressure is itself strictly not a tensor but a scalar: it can *only* act at right angles to the surface (of the boiler). In a similar way surface tension γ has the same units, $mN\,m^{-1}$, as the surface

---

† As there are at least two surface shear moduli for films, the convention G_{ps} will be used for the in-plane shear modulus and G_{ts} for the transverse shear modulus.

stress tensor but can only act in the plane of the surface and at right angles to a small line embedded in the surface. It is for this reason that it is necessary to invoke the kronecker Delta δ. We can see the similarity in argument by taking it a step further. For a monolayer at the surface, the surface pressure π mN m^{-1} is given by:

$$\pi = \gamma_0 - \gamma \qquad (3.4)$$

where γ_0 is the surface tension of pure solvent and γ is the surface tension of the solution.[13] Many workers think erroneously that the occupied fluid/fluid surface is fully characterized physically by the surface tension alone. If this were true, there would be no such thing as surface rheology! Both surface tension and surface stress are usually measured in mN m^{-1} because the Newton itself is such a large unit of force that its use here would be inappropriate. It is an interesting fact that in the older system of units, the cgs system, which was superseded by the SI system some years ago, the units of both surface tension and surface stress were dyne cm^{-1} and that due to a completely fortuitous scaling effect with powers of ten, identical numerical values were obtained in the new SI system of units; for example, the surface tension of water at 21°C is roughly 72 dyne cm^{-1} or 72 mN m^{-1}. The other important point is that due to the presence of extra terms in eqn (3.2) which defines the surface stress in the dilatational experiment as compared with eqn (3.3) which describes the situation for the horizontal plane surface-shear experiment, the subject of surface rheology has evolved historically along two separate paths: surface dilatation and horizontal plane shear.

3.2.2. Surface Strain
The physical components of the surface strain tensor, ε_{ij}, again, if we confine discussions to the horizontal plane, are *four* in number and consist of two normal components and two shear components. As for the bulk, the strain is dimensionless.

3.2.3. Surface Strain Rate
The surface strain rate will be given by $d\varepsilon_{ij}/dt$ *or* $\dot{\varepsilon}_{ij}$ and the units will be reciprocal seconds (s^{-1}).

3.2.4. Surface Elastic Moduli
In the bulk, for an isotropic solid, there are three elastic moduli: Young's modulus (elongational), the shear modulus and the bulk modulus, the latter being essentially an index of compressional resistance under triaxial stress.

In the surface, it would appear that the situation is surprisingly complicated. The reason is that the surface can be deformed in at least three different ways.

   (i) In horizontal plane shear, the plane remaining flat and the directions implicated being 1 and 2.
  (ii) In plane dilatation, the surface remaining flat and the directions implicated being 1 or 2.
 (iii) Mechanical capillary-wave and thermally excited capillary-wave propagation where the surface is subjected to both dilatation and transverse shear. Here the directions implicated are 1 and 3 or 2 and 3. This transverse shear takes place in a vertical plane at right angles to the shear taking place in the horizontal plane in (i) above.

These deformations are illustrated in Fig. 3.1 and the surface is shown populated with an idealized surface-active molecule (hat-pin).

The nomenclature used in surface rheology over the last thirty to forty years is very varied, and although the subject is fairly specialized, there have been about ten active groups throughout the world each with their own set of symbols. Many of them have produced substantial literature.

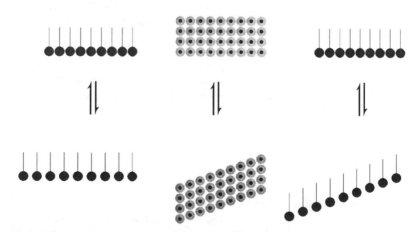

Fig. 3.1. Diagram showing possible surface deformations of a raft of idealized 'hat-pin' molecules. Going from left to right, the three panels correspond to the conditions described in (ii), (i) and (iii) above.

Table 3.1 is included just to illustrate the current confusion and does not claim to be complete.

The two aspects of surface rheology will now be covered in more depth under separate headings.

3.2.5. Surface Dilatational Techniques

At least three different techniques can be identified here which are best classified in order of frequency of disturbance:

    (i) Low frequencies (a few Hz)—oscillating boom Langmuir trough.
   (ii) Medium frequencies (a few hundred Hz)—capillary ripple.
  (iii) Higher frequencies (up to 100 kHz)—thermal ripplon heterodyne spectroscopy.

3.2.5.1. Theory of Dilatational Surface Rheology

The concept of surface dilatational viscosity was probably first introduced by Boussinesq,[14] who studied the rate of fall of a small sphere of liquid in another immiscible liquid. He showed that for the lower hemisphere of the drop, the surface is uniformly expanding from the lower pole, while for the upper hemisphere, the surface is uniformly contracting into the upper pole. Hence, the lower pole acts as a source for new material and the upper pole acts as a sink. (This is quite different from the system studied by Stokes for a rigid sphere moving in a liquid, where, of course, there is no flow of material in the surface of the sphere.) For an expanding surface of an aqueous solution of a surface-active substance, the surface tension will be higher than the equilibrium value, since the surface tension will appear like a physical component of the surface stress tensor. It can be related to the rate of surface dilatation to give the surface dilatational viscosity:

$$\Delta\gamma = K(1/A) \cdot (dA/dt) \qquad (3.5)$$

where

γ = surface tension (mN m^{-1})

K = surface dilatational viscosity (mN s m^{-1})

A = dilating area (m^2)

t = time (s)

If the expansion of the surface of such a solution is suddenly arrested, the surface tension will drop to its equilibrium value.

TABLE 3.1
Nomenclature Used for Surface Rheology by Different Groups

Surface entity	Experimental group (see footnotes)										Symbols[k]
	a	b	c	d	e	f	g	h	i	j	
Surface elasticity											
Elongational *and* dilatational	E_s		ε						Y_s	E, ε	E_s
Horizontal shear				G	G	G^s	G'_s	n, J_s, G_s	G_s		G_{hs}
Transverse shear			γ_0								G_{ts}
Surface viscosity											
Elongational *and* dilatational	η_s	η	ε'								ζ_s
Horizontal shear				η	η	μ^s	$\eta'_s, G''_s/\omega$	η_s			η_{hs}
Transverse shear			γ'								η_{ts}

N.B. The relevant standard nomenclature and physical symbols used should be contained within British Standard BS 5775 (ISO 31/111-1978): as yet, however, surface entities are evidently not covered under British or International Standards. All superior numbers refer to literature references at the end of this Chapter.

(a) A. E. Alexander *et al.*, e.g. *Trans. Faraday Soc.*, 1950, **46**, 235.
(b) E. Dickenson and co-workers.[33,34]
(c) J. C. Earnshaw *et al.*[18,24,25]
(d) D. A. Haydon *et al.*[36]
(e) K. Inokuchi.[37]
(f) R. J. Mannheimer and R. S. Schechter.[32]
(g) P. Sherman and co-workers.[51,52]
(h) E. Shotton and co-workers.[38,39]
(i) N. W. Tschoegl, *J. Colloid Sci.*, 1958, **13**, 500.
(j) Unilever group, Netherlands.[15–17]
(k) Suggested nomenclature and symbols in this review.

The rate of change of surface tension with time will be a function of the bulk phase surfactant concentration and the diffusion coefficient of the surfactant. Clearly it will take a finite time for equilibrium to be reached and the delay in reaching equilibrium can be envisaged in terms of a delayed elastic behaviour. Thus there are two distinct extremes of rheological behaviour in surface dilatation. For a pure liquid, or a solution containing a very soluble solute of high diffusion coefficient, the surface will be ideally elastic. In between these extremes, the surface will be viscoelastic in nature.

3.2.5.2. Low-Frequency Dilatation
The apparatus required for this technique is shown in Fig. 3.2.

The booms B_1 and B_2 have a short sinusoidal excursion of a few millimeters, moving in antiphase so that the entrapped surface may be expanded and contracted symmetrically about the Wilhelmy plate, which senses the surface stress through the surface tension. The technique may be used at the aqueous/air or aqueous/oil interface.

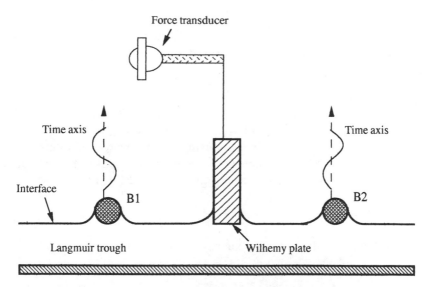

Fig. 3.2. Apparatus for low frequency surface dilatation work, showing the moveable booms B_1 and B_2 and the surface pressure sensing Wilhelmy plate.

The quantity measured is the dynamic surface dilatational modulus (ε) and the theory has been expounded by Lucassen *et al.*[15]

$$\varepsilon = \frac{\mathrm{d}\gamma}{\mathrm{d}\ln A} = |\varepsilon|\exp(i\psi) \tag{3.6}$$

The independent variable in the experiment is the sinusoidal change in the working area:

$$\Delta A = \Delta A_0 \cos \omega t + A$$

where   (3.7)

$$\omega = 2\pi \times \text{working frequency}$$

Owing to delay processes arising from diffusion, the measured response in surface tension will be delayed by a phase angle ϕ. The absolute value of the dilatational modulus $|\varepsilon|$ is approximately equal to $|\Delta\gamma|/|\Delta A/A|$ and the phase angle can also be measured. However, the theory shows that it is only necessary to measure the frequency dependence of $|\varepsilon|$. Measurements of the phase angle would not be mandatory but would be useful to corroborate the results.

For reasons which will become clearer in the next section (capillary ripple) it is useful to define a new parameter such that

$$|\varepsilon| = \varepsilon_0(1 + 2\zeta + \zeta^2)^{\frac{1}{2}} \tag{3.8}$$

ε_0 is the asymptotic value that $|\varepsilon|$ assumes in the limit at high frequencies. Surface chemistry considerations show that:

$$\varepsilon_0 = -\frac{\mathrm{d}\gamma}{\mathrm{d}\ln\Gamma} \equiv \frac{\mathrm{d}\gamma}{\mathrm{d}\ln A} \tag{3.9}$$

where Γ is the surface excess concentration of the surfactant in moles m^{-2}. The reason for this is that at high frequencies we would expect the product ΓA to be constant by the principle of mass conservation of solute. The parameter ζ, conceived during work on capillary ripples, has a value $(\omega_0/\omega)^{\frac{1}{2}}$ where ω_0, a characteristic pulsatance, is defined as

$$\omega_0 = \tfrac{1}{2}D(\mathrm{d}c/\mathrm{d}\Gamma)^2$$

where   (3.10)

$$D = \text{the diffusion coefficient of the surfactant}$$

$$c = \text{the bulk concentration of the surfactant}$$

ω_0 may be calculated for a particular experimental system from the raw ε results *versus* frequency by curve-fitting procedures and may be checked if values of D and Γ are available from other sources. Lucassen *et al.*[15] have demonstrated that the above theory works well not only for single surfactants but also for mixtures of proteins (e.g. poly-L-lysine) and surfactants (sodium dodecyl sulphate). The appropriate value of the diffusion coefficient D is that for the most sluggish species present. Since surface-active chemical substances may interact at the interfaces to produce new species, the method thus provides a powerful analytical tool to investigate such phenomena.

3.2.5.3. Medium Frequencies—Capillary Ripple

As the frequency of the dilatation of a given area of surface is increased it becomes more difficult to maintain the disturbance within a plane surface. Owing to inertial effects, ripples are produced and a complete dynamic description of events requires the mathematics of three dimensions. One of the inevitable effects of this is that the ensuing behaviour of the surface becomes a function not only of the surface film but also the density and shear viscosity of the bulk substrate.

The apparatus required for capillary ripple propagation and measurement on liquid surfaces has been described by Mann and Hansen[16] and is shown in Fig. 3.3.

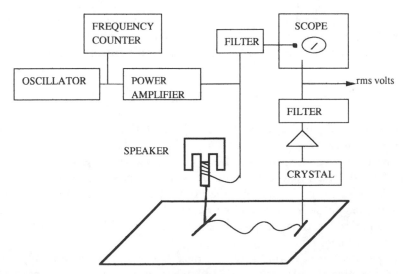

Fig. 3.3. Apparatus for medium frequency surface capillary ripple technique. Transmission of ripples occurs at the moving coil (speaker) drive unit and reception takes place at the crystal transducer.

The experimental raw data available are the ripple wavelength λ and the spatial damping coefficient α. The transmitter consists of a short rod immersed in the surface of the liquid and connected by a vertical wire to an electrical driving transducer (moving coil). The receiver consists of a similar rod exactly parallel to the first, attached to a second electrical transducer (gramophone pickup). An important requirement of the receiving transducer is that its distance from the transmitter can be measured very accurately; also its mechanical inertia should be very low to maximize its sensitivity.

Unfortunately, the presence of the receiver probe in the surface causes it to act as a secondary parasitic transmitter sending out waves which interfere with those received. These effects have to be taken into account in the ensuing analysis. Mann and Hansen[16] introduce β, the attenuation and ϕ the change of phase introduced by the receiver probe itself. Their experiments show that ϕ is effectively zero. For a separation between the detector probe and the source of L m, it is shown that the rms voltage signal appearing at the receiver is

$$\overline{v(L)} = K \exp(-\alpha L) \qquad (3.11)$$

where K is an apparatus constant. If $\beta + \alpha L$ is greater than 3·5, the error in the estimation of α is less than 1%. It also turns out that if L is varied and the phase relationship between the transmitter and receiver is carefully noted, it is possible to obtain the ripple wavelength λ m using the simple relation

$$\lambda = 2\Delta L / \Delta n \qquad (3.12)$$

where Δn is the number of *in phase* and *out of phase* positions noted for a given change ΔL.

The theoretical treatment relating the surface ripple and dilatational rheological behaviours to the bulk density and shear viscosity, diffusion coefficient of the surfactant and the latter's surface excess concentration, was worked out in great detail by van den Tempel and van de Riet.[17] The mathematical treatment is quite extensive and it is only possible to present a summary here.

If we first consider the most complex case, i.e. a surfactant solution having appreciable density ρ and bulk shear viscosity η, the differential equations of motion of a particle in the surface lead to two important boundary conditions. The one for normal stress (referred to as *transverse* stress by Earnshaw[18]) is

$$A(-2\mu k^2 + i(gk\rho/\omega + k^3\sigma_0/\omega - \rho\omega))$$
$$+ B(gk\rho/\omega + k^3\sigma_0/\omega + i(2km\eta)) = 0 \qquad (3.13)$$

and the other for tangential stress (*in-plane* component) is

$$A(-i(2\eta k^2)) + B(-\eta(m^2 + k^2)) - (d\sigma_1/dx)\exp(-ikx) = 0 \quad (3.14)$$

where

A = the complex liquid velocity potential function

B = the complex liquid vorticity function

ρ = bulk density

σ_0 = normal surface stress

$m = k^2 + i\omega/\eta$

$k = 2\pi$(wavelength of ripples)

$\omega = 2\pi$(frequency)

It is readily shown that these conditions immediately simplify to class-ical expressions for special cases, e.g. for an ideal liquid of negligible bulk shear viscosity, the vorticity function B will be zero and the first of the above two conditions becomes:

$$\omega^2 = gk + (\sigma k^3/\rho) \qquad (3.15)$$

which is none other than the Kelvin[19] dispersion equation for waves on a frictionless liquid. The next sophistication is to allow a non-zero bulk viscosity but still exclude the possibility of a surface tension gradient, i.e. no surfactants present. This condition yields two equations of the form

$$a_1 A + b_1 B = 0 \qquad (3.16)$$

$$a_2 A + b_2 B = 0 \qquad (3.17)$$

where a_1, b_1, a_2 and b_2 are complex functions of g, m, η, ρ and σ. The only non-trivial solution eliminating A and B is:

$$a_2 b_1 - a_1 b_2 = 0 \qquad (3.18)$$

this yields the damping relationship for waves:

$$d\ln(\text{wave amplitude})/dx = 2\eta k^3/3\pi\rho f \qquad (3.19)$$

The third and last stage is to include the possibility of a surface-tension gradient ($d\sigma_1/dx \neq 0$), where x is a direction parallel to the surface and in which direction the waves are travelling. Here only approximate solutions are possible.

The theory of van den Tempel and van de Riet has been developed and made more acceptable to the experimentalist by Lucassen and Hansen.[20] The problem here is to deal with both the curvature of the surface produced by the passage of ripples and also with the diffusion coefficient of the surfactant (of diffusion coefficient D m^2 s^{-1}). Fick's Second Law of Diffusion gives

$$\delta \Delta C / \delta t = D \text{ div } \nabla \Delta C_{x,z} \tag{3.20}$$

where z is normal to the surface and x has been defined above. In the numerical system of directions defined for this section, the directions implicated would be, e.g. $\Delta C_{1,3}$. The solution of this equation for a sinusoidal disturbance in the concentration is

$$\Delta C = E \exp(nz) \exp(i(k'x - \omega t)) \tag{3.21}$$

where

ΔC = the deviation of the surface concentration equilibrium

E = a complex constant

$n^2 = (k')^2 - 1\omega/D$

k' = complex ripple wave number = $k + i\alpha$

α = damping coefficient in the x direction

$\quad = d \ln A_z/dx$

A_z = wave amplitude

Writing down the condition for the conservation of transported matter we have

$$\partial(\Gamma A)/\partial t + DA(\partial \Delta C/\partial z)_{z=0} = 0 \tag{3.22}$$

Combination of the three foregoing equations gives the complex dilatational modulus

$$-d\gamma/d \ln A = d\gamma/d \ln \Gamma (1 + i(nD/\omega)(dC/d\Gamma))^{-1} \tag{3.23}$$

If a suitable adsorption isotherm can be found to relate ΔC to $\Delta \gamma$ (the deviation of surface tension from equilibrium) the above expression can

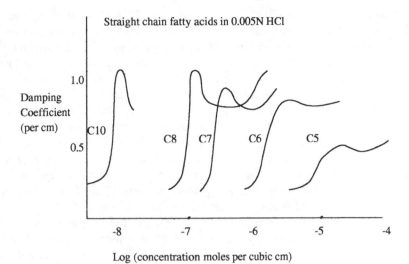

Fig. 3.4. The surface damping coefficient for various straight chain fatty acids in 0·005 N HCl as a function of their bulk concentrations. The symbols C5–C8 & C10 indicate the number of carbon atoms in each aliphatic chain. The results reflect the differences in solubility of each member of the homologous series.

be evaluated completely. Such an adsorption isotherm has been suggested by Szyszkowski[21] and has been used by Lucassen and Hansen.[20] The parameter ζ, the 'transfer coefficient', introduced in the previous section can be related to the complex dilatational modulus in cases where $|\omega/D| \gg |k^2|$. Most experimental results meet this requirement.

The results for many surface-active materials show one or more characteristic maxima in the damping coefficient α when plotted against the concentration C at a fixed operating frequency. This is shown in Fig. 3.4.

As the surface active material becomes less soluble, so the heights of the maxima reduce until for a completely insoluble material, they disappear altogether. They are, of course, absent for 'very soluble materials' and pure liquids.

3.2.5.4. High Frequencies—Thermal Ripplon Heterodyne Spectroscopy
The mechanical inertia of the moving parts of the apparatus prevents the capillary ripple technique to be taken to frequencies higher than a few tens of kHz.

However, Bird and Hills[22] have shown that it is possible to measure the wavelength and the relative intensity of minute ripples (ripplons) that are generated thermally on the surface of any liquid. The amplitude of these ripplons is only a few microns and since their origin is random thermal disturbance, their frequency spectrum is very wide (0–100 kHz). The wavelength of the ripplons is measured as a function of a statistical squared intensity (power spectrum) using a laser light source, reflected from the surface of the liquid (or aqueous/oil interface). The interference of the reflected light rays from the neighbouring parts of the surface of the ripplons is detected using a photomultiplier tube. The output from the photomultiplier tube is then filtered and passed to a real-time autocorrelator. An ensemble of suitable equipment is available from Malvern Instruments Ltd and is shown diagrammatically in Fig. 3.5.

Although there had been growing interest in studying scattered light from the fluid surface during the early 1900s, it waned in the mid 1920s and really had to wait for the invention of the laser before it could be revived.

The relation between surface fluctuations and light scattering for thin membranes was analysed by Kramer.[23] Kramer considers the longwave fluctuations of a thin massless membrane separating two viscous fluids. The intensity and spectrum of the light scattered inelastically from thermal fluctuations is calculated. The interface at equilibrium is considered to lie in the xy plane, with the z axis normal to this plane. In theory there should be *three* characteristic wave frequencies associated with the various modes of deformation of the membrane.

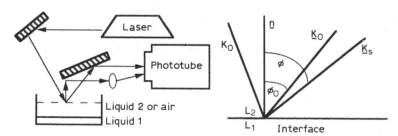

Fig. 3.5. Schematic arrangements for thermal ripplon heterodyne spectroscopy. The lefthand panel shows the optical arrangements for the incident laser light and the subsequent collection of the reflected and scattered light. The righthand panel gives a close-up of the incident ray, K_0, the normal to the working surface $\underline{n}$, the reflected ray $\underline{K}_0$ (reference), and the scattered ray $\underline{K}_s$. The underline indicates **bold type** for the vectors.

For shear, within the xy plane we have:

$$\omega_{ps} = q(S/\Gamma)^{\frac{1}{2}} \qquad (3.24)$$

where $S =$ complex in-plane shear elasticity and for compression, within the xy plane we have

$$\omega_e = q(K/\Gamma)^{\frac{1}{2}} \qquad (3.25)$$

where $K =$ complex in-plane compressional elasticity, whilst transverse to the xy plane we have

$$\omega_T = q(\Pi/\Gamma)^{\frac{1}{2}} \qquad (3.26)$$

where $\Pi =$ complex transverse elasticity.

Unfortunately, as far as light scattering is concerned, the shear modes do not couple to any fluctuations of the dielectric tensor and so cannot be detected by light scattering. This theoretical consideration has, of course, important implications for the experimentalist. However, there is evidence that the shear modes could couple indirectly through mechanical effects.

Any spatio-temporal disturbance in the surface can be described by the equation

$$\zeta(x, t) = \zeta_0 \exp(\mathbf{i}(qx + \omega t)) \qquad (3.27)$$

where

$$\zeta(x, t) = \text{the magnitude of the disturbance}$$

$$\zeta_0 = \text{the amplitude of the disturbance}$$

$$q = \text{the wave number}(2\pi/\Lambda)$$

$$\omega = \text{the pulsatance}$$

Because the waves will behave as a parallel array of ridges and troughs, the system will perform as a weak diffraction grating. Earnshaw[24] describes how the intensity of scattered light I is related to the energy of the surface by the expression

$$\frac{1}{I_0} \frac{dI}{d\Omega} = \frac{\kappa_0^4}{4\pi^2} \{k_B T/(\gamma q^2 + (\rho - \rho')g)\} f \qquad (3.28)$$

where f is a factor containing Fresnel parameters appropriate to the experimental geometry and Ω is the solid angle associated with light-scattering theory (see Ref. 23).

The expression on the right-hand side is also a function of the absolute temperature because it is closely related to the time-average energy of $A \, m^2$ of the surface

$$0 \cdot 5 < \zeta(q, 0)\zeta^*(q, 0) > A(\gamma q^2 + (\rho - \rho')g) \tag{3.29}$$

where ζ^* is the complex conjugate of ζ.

Although ζ is expressed in the time domain in eqn (3.26), using a Fourier transform it can easily be represented in the frequency domain as $\zeta(\omega, t)$. A real-time autocorrelator can be used for the transformation. By the equipartition of thermal energy, the two expressions (3.27) and (3.28) can be combined to give

$$\langle \zeta(q, 0)\zeta^*(q, 0) \rangle = k_B T / (\gamma q^2 + (\rho - \rho')g) \tag{3.30}$$

where $(\rho - \rho^1)$ is the density difference across the interface. Earnshaw et al.[25] explain how the spectrum of scattered light from the surface is close to a lorentzian. The scattered light is characterised by a frequency shift ω_0 and a linewidth Γ_w. It is possible to identify these with the frequency and temporal damping constant of the surface wave.

The expressions for a fluid with only surface tension γ_0 and non-zero bulk shear viscosity η_b are simple to the first order:

$$\omega_0 = (\gamma_0 q^3 / \rho)^{\frac{1}{2}}$$

and   (3.31)

$$\Gamma_w = 2\eta_b q^2 / \rho$$

Initially, the spectrum is convolved with an instrumental function which has to be removed subsequently to reveal the data of interest in the surface system. Earnshaw et al.'s approach[25] is to characterize the fluid surface properties with two complex parameters, the complex surface tension γ and the complex dilatational elasticity ε. The real parts of these parameters relate to the conventional surface tension γ_0 and the compressibility ε_0, while the imaginary parts relate to the transverse shear viscosity γ and the dilatational viscosity κ, respectively.

The observables ω_0 and Γ_w are non-linear functions of γ_0, γ, ε_0 and κ as shown in Fig. 3.6.

Earnshaw has shown, using simulated data and added noise, that whilst γ_0 and ε_0 are relatively immune to noise in the raw data up to noise levels of 10^{-2}, γ and κ become seriously affected in the range 10^{-3} to 10^{-2}. It has been suggested that since the experimental time required to capture the surface data is only of the order of one minute, the technique lends itself to the study of non-stationary biological systems.

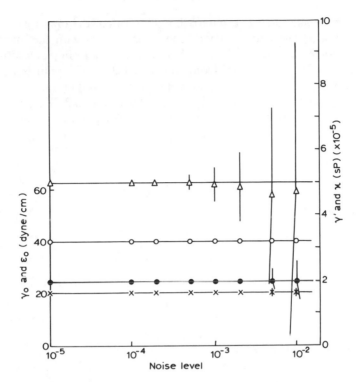

Fig. 3.6. The real and imaginary parts of the complex surface parameters γ and ε as functions of simulated noise (see text).

3.2.6. Surface Shear Rheology

In surface shear rheology (horizontal in-plane shear), rheological measurements are carried out at constant area per molecule of film substance. If a Langmuir Trough is used, the area per molecule may be varied between measurements by moving the boom to different positions. Annular geometry can be preserved by the use of circular guard rings.

3.2.6.1. Continuous Rotation

Classical work on surface shear viscosity (η_{ps}) has been carried out by Fourt and Harkins[26] and Davies.[27] Later, results in many different fields of application have been reported by Joly,[28] Joos,[29] and Carless and Hallworth[30] to mention just a few. Joly and Joos have independently used Ree and Eyring's rate theory[31] to discuss the surface shear viscosity of single species and mixed species monolayers.

As Mannheimer and Schechter[32] have indicated, the only really accurate method of measuring surface shear viscosity is with the improved canal viscometer. The principle of this apparatus is to impart motion to a surface or film of material from the bulk below. The film is trapped between stationary concentric canal walls and the motion imparted from a 'moving floor' at the base of the rheometer (see Fig. 3.7).

The surface in-plane shear viscosity η_{ps} is then given by

$$\eta_{ps} = (y_0 \eta_b / \pi)(t_c / t_c^* - 1) \coth(\pi D) \tag{3.33}$$

where

η_b = bulk shear viscosity of the liquid

y_0 = width of the canal

D = the canal depth–width ratio

t_c = one period of rotation of a particle floating in the film

t_c^* = one period of rotation of a particle floating in pure solvent

It is easily arranged that the ratio of the depth to the width of the canal D is very large, giving $\coth(\pi D) \approx 1$. In spite of the great accuracy and sensitivity obtained, the measurements are expensive in terms of human time and effort. This is because the velocity profile in the surface must be observed from the movement of small pieces of Teflon floating in the surface of the liquid.

Dickinson's group at Leeds[33] has used surface shear viscosity measurements to determine the ageing of films of mixed protein using a Couette-type surface viscometer.[34] No quantitative elasticity results were reported, although surface tension was measured using a Wilhelmy plate.

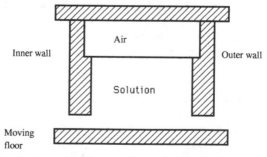

Fig. 3.7. The improved canal surface shear viscometer showing the fixed walls (shaded), the moving floor (shaded) and the solution/air interface.

The work was carried out on various pairs of proteins out of the set: casein, gelatin, lysozyme and α-lactalbumin. These workers confirmed Mussellwhite's earlier observation[35] that casein will replace gelatin at the oil–water interface, if they are both present in solution at the start. However,[33] it had also been discovered that if the interfacial gelatin is allowed to age for several hours at 25°C before introducing the casein, the latter replaced the adsorbed gelatin with great difficulty. Casein on its own, surprisingly, has a low interfacial viscosity at an ionic strength of 0·005 M, although it will replace the gelatin. On the other hand casein has quite a low interfacial tension of about 23 mN m^{-1} against *n*-hexadecane as compared with 35·2 mN m^{-1} for gelatin under comparable conditions. There appears to be very little obvious relation between interfacial tension behaviour and surface shear viscosity behaviour for the same protein.

It is concluded that the surface viscosity of mixed protein solutions is very complex and that while there is often a long time-scale for a single species to reach equilibrium, the time-scale for mixed proteins can be even longer. In contrast to the gelatin, it was found that adsorbed films of lysozyme were less disrupted by the presence of casein. Finally the surface viscosity of all protein films studied proved to be extremely temperature-dependent.

3.2.6.2. Stationary and Non-Stationary Surface Films

Surface dilatational experiments of large disturbance tend on the whole to destroy any kinetic chemical interactions that may occur in the surface. However, these effects could be important to the surface chemist or biologist who would like to use surface rheology as an investigatory tool. On the other hand, surface in-plane shear experiments are less destructive (again at low amplitude) and can lead to the observation of a wide variety of types of film. These surface films may be conveniently classified under four headings:

   (a) stationary: elasticoviscous
   (b) non-stationary: elasticoviscous
   (c) stationary: viscoelastic
   (d) non-stationary: viscoelastic

The term *stationary* needs some explanation before proceeding further. The phenomenon is best stated by writing down the rheological equation of state of a linear elasticoviscous surface film under oscillatory sinusoidal conditions:

$$\eta_{ps}^* = \eta_{ps}' - iG_{ps}'/\omega \tag{3.34}$$

for *stationary films*, this implies

$$\eta_{ps}^*(\omega) = \eta_{ps}'(\omega) - iG_{ps}'(\omega)/\omega \tag{3.35}$$

This means that for a given pulsatance ω, $\eta_{ps}^*(\omega)$ has a unique set of values. For *non-stationary films*, however, we must write

$$\eta_{ps}^*(\omega, t) = \eta_{ps}'(\omega, t) - iG_{ps}'(\omega, t)/\omega \tag{3.36}$$

Now we clearly have to construct the behaviour domain of this function by mapping the real and imaginary parts of η_{ps}^* on to the domain (ω, t). This is non-trivial, however, as ω and t are not altogether independent. There are regions of the universe (ω, t, η_{ps}^*) which are forbidden to the experimentalist, in the sense that the results could be highly inaccurate or meaningless.

Intuitively, it would appear that a necessary restriction on ω is

$$\omega \gg \left[\frac{\partial \eta_{ps}^*(\omega, t)}{\partial t} \right]_\omega \tag{3.37}$$

where the RHS represents a function of the rate of ageing or development of the film. This is because if ω were too low, the film would change properties to a large extent even before the first stress cycle had been executed by the rheometer, leading to a distortion of the output strain sine wave and consequent loss of linearity. A similar argument can be raised for the behaviour of non-stationary viscoelastic films. We will come back and discuss the chemical kinetics of non-stationary films and the implications for surface rheology in a later section.

3.2.6.3. Surface Shear under Constant Stress

Several authors have described techniques for measuring surface shear compliance J_{ps} ($\equiv 1/G_{ps}$) under constant stress. Biswas and Haydon[36] have examined macromolecules including proteins and polysaccharides. Inokuchi[37] has studied proteins. An automated surface rheometer was described by Shotton et al.[38] and used by Kislalioglu et al.[39] to examine thin films of gelatin. A commercial version of the above apparatus was later constructed by Neustadter[40] at British Petroleum, Sunbury, Middx,

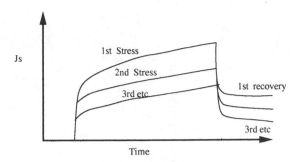

Fig. 3.8. The surface shear compliance (schematic), J_s, for adsorbed *solid* films of many types of soluble polymers. Persistent ageing behaviour makes sensible analysis of viscoelasticity virtually impossible.

UK and manufactured under licence by ZED Instruments at Hersham, Surrey, UK.

It was during work with this technique, that the effect of film growth on the surface compliance was noticed in the case of non-stationary films. The effect is shown in Fig. 3.8.

It can be seen that on successive stressings and recoveries of the film, the scale of the surface compliance gradually diminishes in value. If the film growth is rapid, the usefulness of the constant stress method is severely limited. Also, in its simplest form, the apparatus may not be able to distinguish fully the surface rheological properties from the bulk properties. Oscillation techniques offer some hope of success in overcoming these problems.

3.2.6.4. *Surface Shear Oscillation*

Oscillation techniques (in-plane shear) present a convenient method of studying both stationary and non-stationary films.

Sherriff and Warburton[41] have developed an oscillating ring surface rheometer which can exploit mechanical resonance.[42] The advantage of working near resonance is that the instrument is always near maximum sensitivity.

The apparatus and electronic circuit of the system are given in Figs 3.9 and 3.10, respectively.

When driven with a driving torque T such that

$$T = T_0 \sin \omega t \qquad (3.38)$$

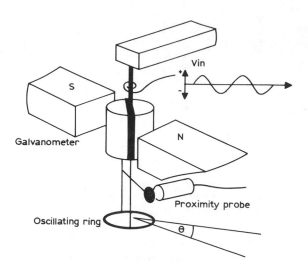

Fig. 3.9. The *normalized resonance oscillating ring surface shear rheometer* (NRORSSR) showing the essential components.

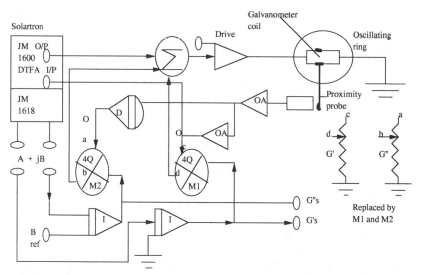

Fig. 3.10. A schematic diagram of the electronic control circuit for the NRORSSR. The right inset shows the original manually controlled potentiometers giving the real and imaginary parts of the surface shear dynamic modulus, later replaced by the automatic control network shown in the main diagram.

where

$$T_0 = \text{amplitude of the torque (Nm)}$$

$$\omega = \text{pulsatance (rad s}^{-1})$$

The behaviour of the surface rheometer is given by:-

$$I\ddot{\theta} + (\phi_1 + \phi_2 + \phi_3)\dot{\theta} + (K_1 + K_2 + K_3)\theta = T \qquad (3.39)$$

where

$I =$ total moment of inertia of the moving parts
$\theta =$ torsional displacement angle about the vertical axis
$\phi_1 =$ internal damping of the instrument
$\phi_2 =$ surface viscous contribution from the sample proportional to η_{ps}
$\phi_3 =$ simulated viscous contribution from the feedback network (for restored resonance: $\phi_2 = -\phi_3$)
$K_1 =$ elastic restoring spring constant for the instrument
$K_2 =$ elastic contribution from the sample, proportional to G'_{ps}
$K_3 =$ simulated elastic contribution from the feedback network (for restored resonance: $K_2 = -K_3$)

As can be seen from eqn (3.39), under conditions of *restored resonance*, the initially perturbed behaviour of the rheometer (due to the influence of the sample) is restored using the feedback network. This functions because ϕ_3 annihilates ϕ_2 and K_3 annihilates K_2. The dynamics of the instrument then revert to those of the instrument alone.

If one wishes to operate at a working frequency either above or below resonance, then the principle of shifted resonance can be used. Using negative feedback ($-K_3$), the working frequency can be moved below natural resonance (down to zero Hz, in theory). Using positive feedback ($+K_3$), the working frequency can be moved above resonance limited by the amplification factor of the feedback amplifier.

The control signal used to calculate the required amount of feedback is derived from the phase conditions at resonance. At *phase* resonance, there is a phase lag of $\pi/2$ radians. Put another way this means that the real part of the output (displacement vector) is zero at phase resonance and hence the quadrature part is exactly the modulus of this vector. This makes the assessment of gain and hence the evaluation of surface viscosity simplistic. Let us see how the control works. If the conditions are slightly away from resonance, then the sign of the real part of the output can be used as a control signal to restore resonance. If the phase angle is slightly less than $\pi/2$, then the real part of the output will have a small negative

value while if the phase angle is slightly more than $\pi/2$ the real part of
the output will have a small positive value. By inspection of Fig. 3.10, it
can be seen that the integrators deployed can close the control loops and
stable control may be achieved. Although viable, this method of control
using integrators is slow and a better method is to attempt to calculate
the change of state required to correct the error. In the general case of
universal systems, this has been solved by Fatmi and Resconi[43] and later
by Sherief and Fatmi,[44] using Lie-like calculus. This method is essential
when studying very rapidly maturing non-stationary films (see cascade
kinetics (Section 3.2.7.3) later): otherwise large errors will arise in the
estimate of surface rheological parameters.

The surface shear elasticity G_{ps} and viscosity η_{ps} can be calculated
immediately from the equilibrium feed-back signals:

$$G_{ps} = \beta_1 C_p C_g / g_f R \qquad (3.40)$$

$$\omega \eta_{ps} = \beta_2 C_p C_g / g_f R \qquad (3.41)$$

where

β_1 and $\beta_2 =$ gain factors for the two feedback amplifiers

$C_p =$ the displacement probe sensitivity ($V\,m^{-1}$)

$C_g =$ the drive coil sensitivity ($m\,N\,A^{-1}$)

$g_f =$ geometry factor for the measuring system
(annular rings) (m^2)

$R =$ reference resistor in the output stage of the
rheometer (Ω)

3.2.7. The Surface Rheology of Non-stationary Surface Films
In 1971,[45] the term *anisotropechrony* was coined to describe non-stationary rheological systems.

The surface rheology of non-stationary surface films is interesting
because the data can be used to study the chemical kinetics of film formation. The systems of interest here are predominantly the macromolecular
polyelectrolytes, which play a crucial role in biological and pharmaceutical processes.

These films may be conveniently classified under three headings.

(i) Liquid films: first-order kinetics.†

---

† The conventional rheological approach is to classify liquids generally as *elasticoviscous* and solids as *viscoelastic*. However the problem of *anisotropechrony*
makes this virtually impossible for ageing surfaces.

(ii) Solid films: second-order kinetics.
(iii) Immunological processes: cascade kinetics.

In general all these molecular systems will behave according to the ambient pH, ionic strength and temperature. These parameters will differ in value between the surface and the bulk.

3.2.7.1. Liquid Films
Liquid surface films often arise when polyelectrolytes are adsorbed at the surface. An example is the lubrication of animal joints by the polymer hyaluronic acid, studied by Kerr and Warburton.[46,47] The graphic formula is shown in Fig. 3.11.

It will be noticed that the charge density along the spine of the molecule is very high when fully ionized, due to the presence of alternating D-glucuronic acid residues. The pK_a of D-glucuronic acid is of the order of 4 and according to the extended Henderson–Hasselbach equation, the charge density drops to half its maximum value at the equivalent pH, i.e. at pH 4. However the pH in healthy non-inflamed joints could be expected to be close to 6·9 and under these conditions, the hyaluronic acid molecule will be almost fully ionized.

An ensemble of highly charged molecular spheres will not show any overlapping of sphere boundaries, due to Coulombic repulsions between the spheres. Thus the lubricative, and perhaps more importantly the load-bearing, properties of hyaluronic acid films appear to arise from these

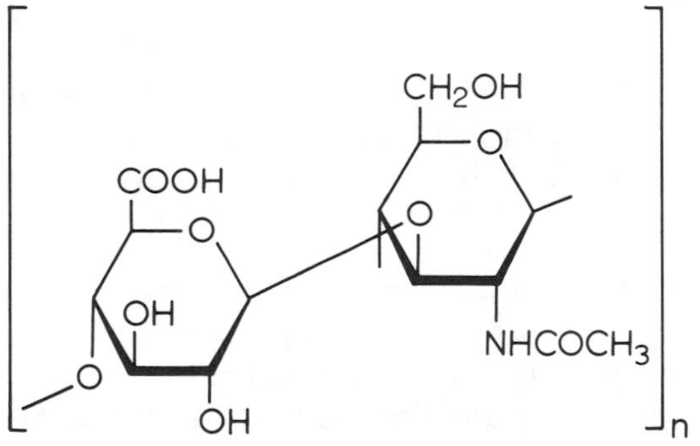

Fig. 3.11. Simplified graphical formula for the repeating unit in hyaluronic acid. The lefthand part of the structure represents D-glucuronic acid which is charged when the ambient pH is physiological or above. Under healthy conditions this gives rise to a lubricating ensemble of 'micro ball bearings' in the human and animal joint.

endowments. The healthy animal joint in some respects appears to be a miniature ball-race! Certainly no engineer has yet built a liquid lubricant bearing with a lower coefficient of static or moving friction or greater load-bearing capacity.

In rheumatic and related disease, it is known that locally in the region of the diseased cartilage and synovia, the local pH can drop by one or two whole pH units due to a high turnover of lactic acid.[48] Kerr and Warburton[46,47] using the normalized resonance oscillating-ring surface shear rheometer (NRORSSR), found that in the region of physiological pH, all the hyaluronic acid solutions studied exhibited only measurable surface viscosity with virtually zero surface elasticity. Some results are shown in Fig. 3.12, reproduced with permission of Dr Helen Kerr.

As the pH is lowered to mildly acid conditions the strength of the film falls off dramatically and towards pH 4 (3·7) the surface elasticity had reached an alarming 25 mN m^{-1} after only 2 min. One could imagine that in chronically inflamed rheumatic disease what was once an ideal lubricant has now become a superglue cement!

Kinetics of formation of liquid films. The liquidity of these films reflects the absence of strong second-order bonds of energy greater than $k_B T$ at ambient temperature. We have seen that in the case of hyaluronic acid, the essentially non-entangling molecules are kept in isolation by means of their high charge density. They will, of course, be highly hydrated at the same time.

Inspection of Fig. 3.12 shows a characteristic type of behaviour for all solutions which may be summarized as follows:

(1) All solutions exhibit anisotropechrony with respect to surface viscosity, often taking several minutes to reach equilibrium.
(2) The half-life for the total change and the extent of total change depend upon the bulk concentration of hyaluronic acid and the presiding pH.

There is evidence that on arrival at the surface, the polyanions of hyaluronic acid will witness a step-change lowering of dielectric constant. The dielectric constant of water is approximately 80, but that for air or vacuum is 1, whilst for many lipids it is about 2. Coulomb's law states that the force between coulombic charges is

$$F = \frac{q_1 q_2}{\varepsilon_d r_{12}^2} \tag{3.42}$$

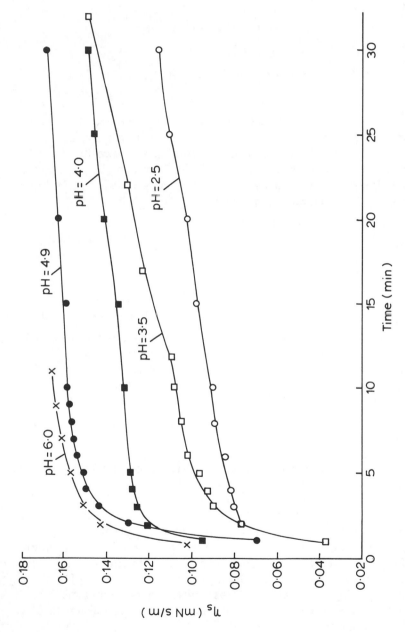

Fig. 3.12. Surface viscosity *versus* time for solutions of hyaluronic acid showing the effect of varying bulk phase pH (solvent = 0·2 M phosphate, 0·145 M NaCl buffer, conc. = 4·0 mg · ml^{-1}, f = 6·0, · = 20 ± 1°C). (Kerr and Warburton, 1985).

where

$$F = \text{force of repulsion or attraction}$$

$$q_i = \text{coulombic charge } (+ \text{ or } -)$$

$$r_{12} = \text{distance of separation of charges}$$

$$\varepsilon_d = \text{dielectric constant}$$

Thus, a step-change of dielectric constant will provoke a step-change of force acting within the polyion in the case of hyaluronic acid.

The data in Fig. 3.12 shows clearly that the surface viscosity of hyaluronic acid increases on forming the film. This increase implies an increase in the hydrodynamic volume as given by Einstein's equation for a dilute suspension of non-interacting spheres:

$$\eta = \eta_0(1 + 2 \cdot 5\phi + \text{higher terms}) \qquad (3.43)$$

where

$$\eta_0 = \text{viscosity of the vehicle}$$

$$\eta = \text{viscosity of the suspension}$$

$$\phi = \text{volume fraction of the suspended material}$$

In the system considered, if the hydrodynamic volume of a single molecule is v_m μm^3 and the concentration of the solution is C moles litre^{-1} the value of ϕ will be $NCv_m \cdot 10^{-15}$ where $N = $ Avogadro's number. The rate of increase of surface viscosity with time will therefore be a first-order kinetic process:

$$\eta_{ps,t} = \eta_{ps,\infty}(1 - \exp(-\lambda t)) \qquad (3.44)$$

where

$$\eta_{ps,\infty} = \text{equilibrium surface viscosity at long times}$$

$$\lambda = \text{time constant}$$

All other things being equal, the value of λ will depend finally upon the degree of ionization α of the polyelectrolyte. If the solution is acidic (as with rheumatic disease), α will be low and hence λ will be low, leading to a long time-scale for the establishment of a lubricating film (early-morning stiffness). On the other hand, in healthy joints the pH will be close to physiological, α will be almost unity, giving the maximum possible value to λ and hence the minimum time for the establishment of a good lubricating film.

Two important points arise here.

(i) Under very acid conditions, the degree of ionization of the poly-electrolyte may be so low as to cause a fundamental change of rheological behaviour, i.e. from liquid film to solid film.

(ii) Many treatments for rheumatic illness and stiff joints imply the local raising of pH, i.e. the implicit use of a surface buffer.

For example, Indomethacin Sodium is the sodium salt of a weak acid: it is known to be concentrated in synovial fluid after treatment. Another older remedy for stiff joints is White Liniment.[49] This is a water-in-oil emulsion containing *alkaline* ammonium chloride solution, in addition to the emulsifier, ammonium oleate.

Other examples of polymeric systems giving liquid films arise from solutions of sodium alginate and also sodium acrylate. On the other hand, many solutions of lower molecular weight but highly charged surfactants, e.g. sodium dodecyl sulphate and cetyl pyridinium chloride, will also give rise to liquid films in non-biological studies.

3.2.7.2. Solid Films

Solid films often arise when polyelectrolytes of low charge density get adsorbed at the liquid surface. Many proteins or polysaccharides covalently bound to proteins (e.g. gum acacia senegal) will give rise to such films.

Again, these films do not arise at full strength spontaneously by adsorption: rather they take tens of minutes or even hours to reach equilibrium. The rate of formation of collagen films[50] has been studied over a range of pH. The surface viscosity of these films was low and their time-behaviour complicated, sometimes increasing and sometimes decreasing. However, the development of the surface shear elasticity G_{ps} exhibited consistent and characteristic behaviour. It was also very pH-dependent.

A useful data transformation, the origin of which will be explained in the next section, is to plot the reciprocal of G_{ps} against the reciprocal of time.

For a fully adsorbed film which then ages due to the formation of intermolecular cross-links, the transformed data points would be expected to lie on a straight line. The gradient and intercept of this line on the $1/G_{ps}$ axis would then be related to the type of adsorbed polymer, its bulk concentration, the ionic strength and prevailing pH.

Some results for the protein collagen (0.003% (w/v)) taken from a previous publication[50] are given in Figs 3.13 and 3.14 over a wide range of pH.

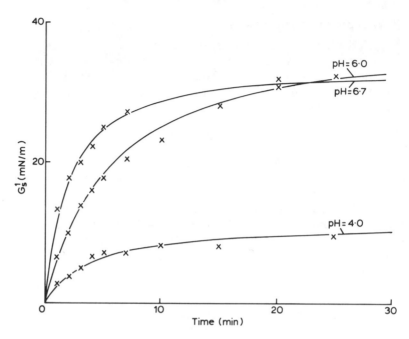

Fig. 3.13. The real part of the surface shear dynamic modulus for 0.003% w/v collagen (Gattefosse reagent Pancogine) as a function of time for the pH range 4·0–6·7. The solid lines are the least-sum-of-squares-of-errors fit to eqn (3.46) transposed back to suit the raw data.

Since then, many other similar systems have been studied. Martinez-Mendoza and Sherman[51] and Reeve and Sherman[52] have studied the formation of protein–monoglyceride and protein–diglyceride complexes across the aqueous/oil interface using oscillatory surface shear rheometry. These results are of particular interest in the food industry for the design of new types of food product.

Buhaenko *et al.*[53] have looked at the surface shear rheology of mono-layer deposited films of long-chain fatty acids and their divalent metal soaps. In particular, docosanoic acid was investigated over a range of surface pressures and pHs. The dynamic surface shear modulus was found to be very dependent upon surface pressure above $\pi = 20$ mN m^{-1}, whilst the surface shear elasticity reached a maximum at pH 9·5 (80–100 mN m^{-1}) and minimum at pH 3·5. It was conjectured that the latter result arose due to structuring of the film imposed by ions in the bulk phase. Most encouraging results were produced for the prediction of Y-

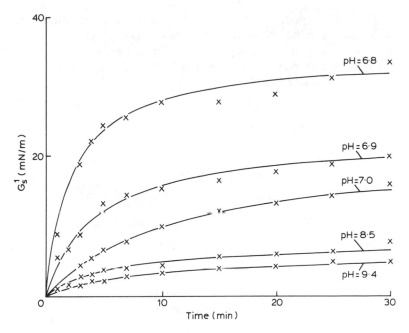

Fig. 3.14. Continuation of the results series shown above in Fig. 3.13 in the pH range 6·8–9·4.

type deposition of monolayers of the fatty acids in the presence of divalent cations of the type Cu^{++} and Cd^{++} from surface rheological data. These deposited films represent a rejuvenated interest in Langmuir–Blodgett technology.

Carlin *et al.*[54] have studied gum acacia senegal and found that interfacial films age almost indefinitely, a fact that was indicated from earlier work.[55]

Kinetics of formation of solid films. The kinetics of formation of solid films at the solution interface can be considered under various degrees of elaboration.

(i) The simplest case is for a single-component fully adsorbed species. This was first published in 1978,[50] and the reader is referred to the original text for the full derivation.

The essential steps are as follows. We first make the assumption that all the solute species are adsorbed at the surface. For even large polymer molecules this is apparently true after five to ten minutes, although in-plane kinetics can be active for many hours due to the low in-plane diffusion caused by a locally highly viscous environment.

The rate of coagulation of small particles in an aerosol or fog was derived by Smolouchowski:[56]

$$\left(\frac{dn}{dt}\right)_- = -n^2 4\pi D_s R \tag{3.45}$$

where

$$n = \text{no. of free particles}$$

$$D_s = \text{in-plane diffusion coefficient}$$

$$R = \text{radius of capture for coagulation}$$

(The negative sign by the differential indicates a negative rate of change of particles with time.)

We can apply this concept to a thin molecular multilayer at the surface and treat neighbouring flexible polymer segments as small particles. After coalescence, cross-links (non-covalent, second-order) will form a gel network.

Using very simple rubber elasticity theory,[57] we can write:

$$G'_{ps} = kn_i = k(n_0 - n_t) \tag{3.46}$$

where

$$k = \text{a constant for isothermal conditions}$$

$$n_i = \text{no. of immobilized segments}$$

$$n_0 = \text{no. of initial segments}$$

$$n_t = \text{no. of segments at time } t$$

Incorporation of eqn (3.45) into (3.46) yields:

$$1/G'_{ps}(t) = 1/G'_{ps}(\infty) + \frac{1}{at} \tag{3.47}$$

where

$$a = 4\pi D_s R n_0$$

For very dilute solutions of some polymers, this so-called double reciprocal plot gives quite an accurate model of the surface coagulation kinetics. However, the relationship breaks down at high bulk concentrations of adsorbing species especially if the latter do not adsorb well at low bulk concentrations. Under these conditions, we need the second elaboration.

(ii) The second case is the single component continually adsorbing into a surface sink. This case can be graphically illustrated with an analogy, although analogies are often dangerous!

The author remembers as a child during the second world war that the local grocer's store used to keep down the number of flies especially during the summer months with a fly-paper. This was before the widespread use of insecticides or UV lamps for this purpose. The surface of the fly-paper was very sticky and the paper itself was about 4–5 cm wide and half a metre long. Any hapless flies landing on the sticky surface stayed there for good. However, once a monolayer of flies had formed, the fly-paper ceased to work because the flies themselves were not sticky! The analogy works for the formation of a molecular monolayer when the forces of attraction are arrested at this stage. However, molecules are not flies, and it is quite possible for the first molecular layer to adsorb further layers leading to a multilayer.

The amount of material entering the surface will be given by Fick's First Law:

$$J = -D_B \nabla C \tag{3.48}$$

where

$$J = \text{material flux (moles cm}^{-2}\,\text{s}^{-1})$$

$$C = \text{bulk concentration}$$

$$D_B = \text{bulk diffusion coefficient}$$

When this influx is included, it will modify eqn (3.45). Equation (3.48) may be rewritten expressing J in terms of dn/dt

$$J = J_0 \frac{dn}{dt}$$

$$\left(\frac{dn}{dt}\right)_+ = -\frac{D_B}{J_0} \nabla C = k_0^2 4\pi D_s R \tag{3.49}$$

Under the present conditions ∇C will be negative and constant to a first-order approximation, especially if the coagulation process itself forms a perfect flux sink. Combining eqns (3.45) and (3.49) we can write the net

dn/dt as the sum of negative and positive processes.

$$\frac{dn}{dt} = (k_0^2 - n^2) 4\pi D_s R \qquad (3.50)$$

Expressing this as an integral, we have

$$\int \frac{dn}{k_0^2 - n^2} = 4\pi D_s R \, dt = a_0 \, dt \qquad (3.51)$$

According to Gradsteyn and Ryzhik,[58] the LHS is a standard integral, the solution of which is

$$\frac{1}{2k_0} \ln \left(\frac{k_0 + n}{k_0 - n} \right) \qquad (3.52)$$

On integration of the RHS also,

$$\frac{1}{2k_0} \ln \left(\frac{k_0 + n}{k_0 - n} \right) = a_0 t + c_I \qquad (3.53)$$

where c_I = integration constant.
When $t = 0$, $n = n_0$, so

$$c_I = \frac{1}{2k_0} \ln \left(\frac{k_0 + n_0}{k_0 - n_0} \right)$$

Finally

$$\frac{1}{2k_0} \ln \left[\left(\frac{k_0 + n}{k_0 - n} \right) \left(\frac{k_0 - n_0}{k_0 + n_0} \right) \right] = a_0 t \qquad (3.54)$$

Again, we can insert eqn (3.45) into this to obtain the real part of the shear modulus G'_{ps}.

Qualitatively, the difference between system (i) and system (ii) is shown in Fig. 3.15.

It will be noticed that system (i) will eventually asymptotically reach equilibrium at $G'_{ps,\infty}$. However, system (ii) cannot reach equilibrium due to the continual influx of new material.

Carlin et al.[59,60] have studied aqueous solutions of the complex glycoprotein gum acacia senegal and concluded that its behaviour is best classified under system (ii).

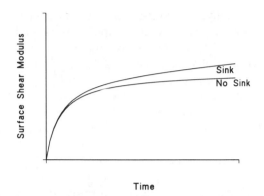

Fig. 3.15. The real part of the surface shear dynamic modulus as a function of time. The two curves are theoretical evaluations of (i) eqn (3.47) (No Sink) and (ii) eqn (3.54) solved with eqn (3.45) (Sink).

(iii) Heterogeneous second order kinetics.

If two different solutes A_1 and A_2 coagulate at the surface, we have three possibilities:

(a) $n_1 + n_1 \to 2n_1$
(b) $n_2 + n_2 \to 2n_2$
(c) $n_1 + n_2 \to n_1 n_2$

where n_1 and n_2 represent numbers of molecules of species A_1 and A_2. Although preliminary experimental work has been published for acacia senegal in the presence of second components,[61] no kinetic models have yet been analytically determined.

3.2.7.3. *Immunological Processes: Cascade Kinetics*
Perhaps one of the most exciting growth areas of surface rheology from the biological viewpoint is the quantitative monitoring of immunological reactions. An example of such a system is the activation of a pro-clotting enzyme and its subsequent cleavage of the protein coagulogen, studied by Nithiananthan[7] using the NRORSSR.

Under certain circumstances, a pro-clotting enzyme of molecular weight 150 000 may be activated by scission to give an active enzyme of molecular weight 84 000. The active enzyme will then cleave the coagulogen molecule into three moieties A, B and C, C being soluble.

Autoxidation of sulphydryl groups contained on the fragments A and B then cause the rapid formation of a disulphide cross-linked gel. The overall process is depicted schematically in Fig. 3.16 whilst a typical set of surface rheological results are shown in Fig. 3.17. These illustrations are reproduced by permission of Ahilan Nithiananthan.

An important principle here is that the reactants are concentrated by several orders of magnitude by virtue of being surface active. This allows the kinetics of the reaction to be monitored even at low concentrations of reactants. It will be noticed that at early stages the surface shear elasticity is low or non-existent, reflecting the absence of gel cross-links at early times.

However, this is just one example of a primitive immunological process. The application to human immunological processes and the subsequent development of sensitive diagnostic techniques is yet to be explored.

3.3. SUMMARY AND CONCLUSIONS

Surface rheology has come a long way since the 1920s, the most dramatic progress being made in the last ten years. It is clear that its progress was held up until the invention of the laser, the personal computer (PC) and specialist devices such as the NRORSSR. On the biological side it had to wait for the most recent understanding of the physical chemistry and thermodynamics of polymer and surfactant adsorption.

It is a more subtle technique than bulk rheology, especially due to its occasional functional dependence on surface anisotropechrony. In medicine, particularly, it has great potential as an extra diagnostic technique in arthritic disease, and immunologically linked conditions. In dentistry and orthodontics, it has useful application in the development of saliva substitutes and the concomitant understanding of the biophysics of natural saliva.

In industries such as pharmaceutical, agrochemical and food, its potential lies in the study of emulsion stability and quantification of lipid–protein interactions.[51,52] In the paint industries, it can easily monitor the surface drying of paints in the presence of environmentally friendly metals replacing toxic lead.

Lastly, in complex systems and formulations it seriously undermines the naive adoption of surface tension alone as a criterion for wetting.

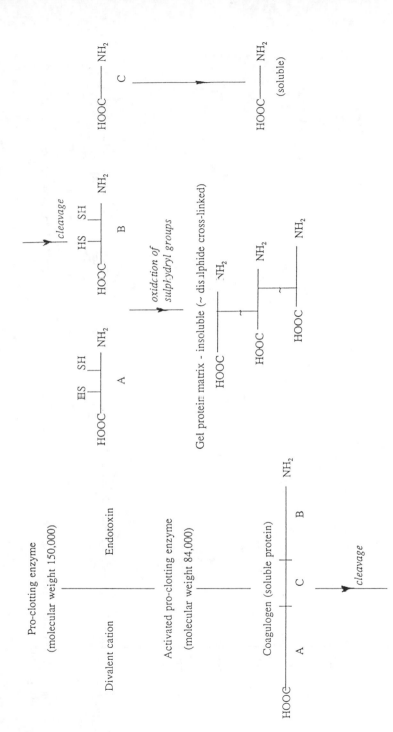

Fig. 3.16. Stages in the cleavage, oxidation and subsequent cross-linking of a special coagulogen protein in the presence of a proclotting enzyme and traces of endotoxin. The reactants are concentrated at the aqueous air/interface giving enhanced detection by surface rheology.

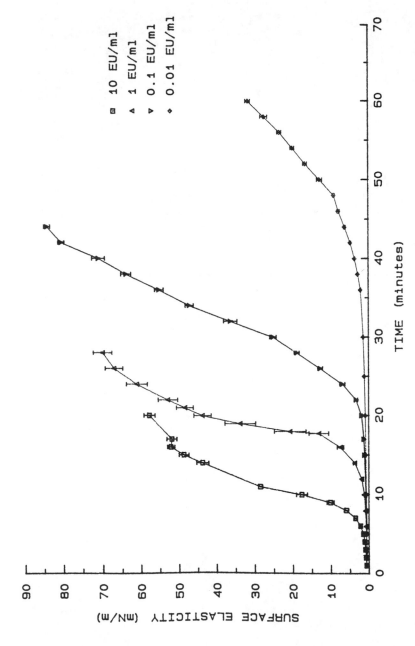

Fig. 3.17. The reproducibility of the surface elastic film formed during the reaction of 10% LAL with different concentrations of endotoxin.

ACKNOWLEDGEMENTS

The author acknowledges with gratitude the support and interest of senior colleagues and PhD students past and present at the School of Pharmacy, University of London over many years. Also, particular thanks are due to Prof. John Earnshaw of the Department of Physics, Queen's University, Belfast, Northern Ireland for time on his thermal ripplon apparatus and helpful discussions. In addition, the author is grateful to Mr Stuart Smith, Lecturer in Orthodontics at the Royal London Hospital School of Dentistry, for help with diagrams produced on the Apple Mac PC and to Dr Michael Moody, Reader in Pharmaceutical Chemistry at the School of Pharmacy, University of London, for help with the solution of eqn (3.50).

Finally, I must thank Dr Anthony Collyer of the School of Sciences, Sheffield Hallam University for the invitation to contribute to this important book.

NAMES AND ADDRESSES OF INSTRUMENT MANUFACTURERS

Malvern Instruments, Spring Lane, Malvern, Worcestershire, UK, WR14 1AQ.
Surface Science Enterprises, 37A Oatlands Avenue, Weybridge, Surrey, UK, KT13 9SS.
ZED Instruments Ltd, 336 Molesey Road, Hersham, Surrey, UK, KT12 3PD.

REFERENCES

1. B. Warburton, 'Whither Rheology in Pharmacy?', *J. Texture Studies*, 1970, **1**, 379.
2. B. R. Bird, in *Dynamics of Polymeric Liquids. Vol. 2*, R. B. Bird, O. Hassager, R. C. Armstrong and C. F. Curtis (Eds), John Wiley & Sons, New York, 1977.
3. A. Silberberg, *J. Phys. Chem.*, 1962, **66**, 1872.
4. A. Silberberg, *J. Chem. Phys.*, 1968, **48**, 2835.
5. A. Silberberg, in *Ions in Macromolecular and Biological Systems*, D. H. Everett and B. Vincent (Eds), Scientechnica, Bristol, 1978, p. 1.
6. J. F. Danielli, *Nature*, 1945, **156**, 468.
7. A. Nithiananthan, New techniques for the detection of the limulus amoebocyte lysate endpoint. PhD thesis, University of London, 1991.

8. H. R. Kerr, The surface properties of Hyaluronic Acid Solutions in relation to joint lubrication. PhD thesis, University of London, 1985.
9. D. M. H. Barnes, The surface rheological assessment of patients with Sjogren's syndrome, PhD thesis, University of London, 1988.
10. Sir Geoffrey Taylor, in *Proc. of the Second International Congress on Rheology: Oxford 6–30 July*, Harrison (Ed.), Butterworth, 1954, p. 1.
11. J. S. Vermaak, C. W. Mays and D. Kullman-Wilsdorf, *Surface Science*, 1968, **12**(2), 128.
12. P. Sherman, *Industrial Rheology*, Academic Press, 1970, p. 3.
13. M. J. Jaycock and G. D. Parfitt, *Chemistry of Interfaces*, Ellis Horwood, Chichester, 1981, p. 106.
14. J. Boussinesq, *Compt. Rend.*, 1913, **156**, 983, 1035, 1124.
15. J. Lucassen, F. Holloway and J. H. Buckingham, *J. Colloid and Interface Science*, 1978, **67**(3), 432.
16. J. A. Mann and R. S. Hansen, *J. Colloid Sci.*, 1963, **18**, 805.
17. M. van den Tempel & R. P. van de Riet, *J. Chem. Phys.*, 1965, **42**(8), 2769.
18. J. C. Earnshaw, *Biophysical J.*, 1989, **55**, 1017.
19. Sir William Thomson (Lord Kelvin), *Phil. Mag.*, 1871, **42**, 368.
20. J. Lucassen and R. S. Hansen, *J. Colloid Sci.*, 1967, **23**, 319.
21. B. von Szyszkowski, *Z. Physk. Chem. (Leipzig)*, 1925, **116**, 466.
22. M. Bird and G. Hills, in *Physiochemical Hydrodynamics in Commemoration of V. G. Levich*, D. Brian Spalding (Ed.), Advance Publications Ltd, London, 1977, 609–25.
23. L. Kramer, *J. Chem. Phys.* 1971, **55**, 2097.
24. J. C. Earnshaw, in *Fluid Interfacial Phenomena*, C. A. Croxton (Ed.), John Wiley & Sons, New York, 1986, p. 437.
25. J. C. Earnshaw, R. C. McGivern and P. J. Winch, *J. Phys. France*, 1988, **49**, 1271.
26. L. Fourt and W. D. Harkins, *J. Phys. Chem.*, 1938, **42**, 897.
27. J. T. Davies, *Biochem. J.*, 1954, **56**, 509.
28. M. Joly, in *Proceedings of Fifth International Congress on Rheology: Tokyo, 7–11 October, 1968*, Shigehan Onogi (Ed.), University of Tokyo Press, Tokyo, 1968, 191–206.
29. P. Joos, *Rheol. Acta*, 1971, **10**, 138.
30. J. E. Carless and G. Hallworth, *Chemistry and Industry*, 1966, **30**, 37.
31. T. Ree and H. Eyring, *J. Appl. Phys.*, 1955, **26**, 793.
32. R. J. Mannheimer and R. S. Schechter, *J. Coll. & Interface Sci.*, 1969, **32**(2), 195.
33. J. Castle, E. Dickinson, B. S. Murray and G. Stainsby, in Proteins at interfaces: Physicochemical and biochemical studies, ACS Symposium Series No. 343, J. L. Brash and T. A. Horbett (Eds), American Chemical Society, 1987.
34. E. Dickinson, B. S. Murray and G. Stainsby, *J. Coll. & Interface Sci.*, 1985, **106**, 259.
35. P. R. Mussellwhite, *J. Coll. & Interface Sci.*, 1966, **21**, 99.
36. B. Biswas and D. A. Haydon, *Proc. Roy. Soc.*, 1962, **A217**, 296.
37. K. Inokuchi, *Bull. of the Chem. Soc. of Japan*, 1953, **26**(9), 500.

38. E. Shotton, K. Wibberley, B. Warburton, S. S. Davis and P. L. Finlay, *Rheol. Acta*, 1971, **10**, 142.
39. S. Kislalioglu, E. Shotton, S. S. Davis & B. Warburton, *Rheol. Acta*, 1971, **10**, 158.
40. D. M. Grist, E. L. Neustadter and K. P. Whittingham, *J. Canadian Petroleum*, 1981, **20**, 74.
41. M. Sherriff and B. Warburton, *Polymer*, 1974, **15**, 253.
42. M. Sherriff and B. Warburton, in *Theoretical Rheology*, J. Hutton, J. R. A. Pearson and K. Walters (Eds), Interscience, 1975, p. 299.
43. H. A. Fatmi and G. Resconi, *Nuovo Cimento*, 1988, **101B**(2), 239.
44. H. T. Sherief and H. A. Fatmi in *Proc. 2nd Int. Conf. on Microelectronics for Neural Networks*, U. Ramacher, U. Ruckert and J. A. Nossek (Eds.), Kyriell & Method, 1991, p. 193.
45. B. Warburton, *Rheol. Abstracts*, 1971, **14**, 57.
46. H. R. Kerr and B. Warburton, *J. Pharm. Pharmacol.*, 1980, **32**, 57P.
47. H. R. Kerr and B. Warburton, *Biorheology*, 1984, **22**, 133.
48. K. H. Falchuk, E. J. Goetze and J. P. Kulka, *Am. J. Med.*, 1970, **49**, 223.
49. *The Pharmaceutical Codex*, The Pharmaceutical Press, London, 1979, p. 973.
50. B. Warburton, in *Ions and Macromolecules in Biological Systems*, 1978, D. H. Everett and B. Vincent (Eds), Scientechnica, Bristol, 1978, p. 273.
51. A. Martinez-Mendoza and P. Sherman, *J. Dispersion Sci. Technol.*, 1990, **11**, 347.
52. M. J. Reeve and P. Sherman, *Colloid Polym. Sci.*, 1988, **266**, 930.
53. M. R. Buhaenko, J. W. Goodwin and R. M. Richardson, *Thin Solid Films*, 1988, **159**, 171.
54. B. A. Carlin, J. N. C. Healey and B. Warburton, *J. Pharm. Pharmacol.*, 1989, **41**, 105P.
55. K. Wibberley, *J. Pharm. Pharmacol.*, 1962, **14**, 87T.
56. M. von Smolouchowski, *Z. Physik. Chem.*, 1917, **92**, 129.
57. J. D. Ferry, *Viscoelastic Properties of Polymers*, 1980, John Wiley, London and New York.
58. I. S. Gradsteyn and I. M. Ryzhik, in *Table of Integrals, Series, and Products*, Yu. V. Geronimus and M. Yu. Tseytlin (Eds), Academic Press, London & New York, 1965.
59. B. A. Carlin, J. N. C. Healey and B. Warburton, *J. Pharm. Pharmacol.* 1990, **42**, 24P.
60. B. A. Carlin, J. N. C. Healey, B. Warburton, C. Nugent, C. Hughes and J. C. Earnshaw, *J. Pharm. Pharmacol.*, 1991, **43**, 74P.
61. C. A. Moules and B. Warburton in *Rheology of Food, Pharmaceutical and Biological Materials with General Rheology*, R. E. Carter (Ed.), Elsevier Applied Science, 1990, pp. 211–18.

Chapter 4

Large-Amplitude Oscillatory Shear

A. Jeffrey Giacomin

Department of Mechanical Engineering, Texas A & M University, Texas, USA

AND

John M. Dealy

Department of Chemical Engineering, McGill University, Montreal, Canada

4.1. INTRODUCTION

Although linear viscoelastic properties can be measured in many ways, the small-amplitude oscillatory shear test is the most widely used method. Nonlinear viscoelastic properties can also be measured in many ways, but no predominant test method has emerged amongst experimentalists. Whereas there is a unifying theory that describes linear behavior, there is no unifying constitutive theory for nonlinear viscoelasticity. For this reason, each nonlinear test reveals a different aspect of a material's behavior. Hence, experimentalists have designed various transient experiments to capture different features of nonlinear viscoelasticity.

Since oscillatory shear testing allows the strain magnitude and the time scale to be varied independently, large-amplitude oscillatory shear testing is of interest for studying nonlinear viscoelasticity in polymeric liquids. Because oscillatory shear does not involve any sudden jumps in speed or position, it is a relatively easy flow to generate. Error analysis is straightforward for this strain history.

4.1.1. Simple Shear

Simple shear is illustrated in Fig. 4.1 as the flow generated between parallel sliding surfaces with a constant gap. It is easily shown that for uniform fluid properties and in the absence of inertia, the shear rate is uniform throughout the fluid. The shear strain, γ, is defined as the plate displacement per unit sample thickness, the shear stress, σ, is the tangential force per unit contact area, and the shear rate is the time rate of change of the shear strain, $d\gamma/dt$.

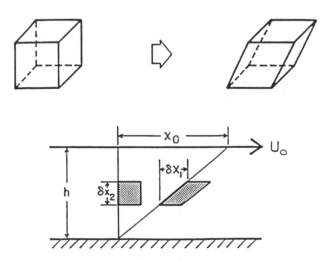

Fig. 4.1. Illustration of simple shear.

4.1.2. Oscillatory Shear

Oscillatory shear is a uniform simple shear deformation in which the shear strain is

$$\gamma(t) = \gamma_0 \sin(\omega t) \tag{4.1}$$

and the shear rate is

$$\dot{\gamma}(t) = \gamma_0 \omega \cos(\omega t)$$
$$= \dot{\gamma}_0 \cos(\omega t) \tag{4.2}$$

where γ_0 is the shear strain amplitude and ω is the angular frequency.

4.1.3. Linear Viscoelasticity

For polymeric liquids, experimental evidence overwhelmingly confirms that when $\gamma_0 \ll 1$, the shear-stress response to oscillatory shear is also sinusoidal.

$$\sigma(t) = \sigma_0 \sin(\omega t + \delta) \tag{4.3}$$

where $\delta(\omega)$ is called the mechanical loss angle, and the amplitude ratio, σ_0/γ_0, is the dynamic modulus:

$$G_d \equiv \sigma_0/\gamma_0 \tag{4.4}$$

For small strains both G_d and $\delta(\omega)$ depend upon frequency only, and not on γ_0. Since σ_0 is proportional to γ_0, this behavior is called linear viscoelasticity. Linear properties are usually reported in terms of the storage and loss moduli which are

$$G' \equiv \sigma_0/\gamma_0 \cos(\delta) \tag{4.5a}$$

and

$$G'' \equiv \sigma_0/\gamma_0 \sin(\delta) \tag{4.5b}$$

This small-strain behavior is described adequately by the theory of linear viscoelasticity, which can be expressed in terms of the Boltzmann superposition principle:

$$\sigma(t) = \int\limits_{-\infty}^{t} G(t-t')\dot{\gamma}(t')\,\mathrm{d}t' \tag{4.6}$$

where $\boldsymbol{\sigma}$ is the extra stress tensor and $\dot{\boldsymbol{\gamma}}$ is the rate of deformation tensor. This show that, within the linear theory of viscoelasticity, the response to any deformation history can be calculated if the relaxation modulus, $G(s)$, is known. For simple shear

$$\dot{\gamma} = \begin{bmatrix} 0 & \dot{\gamma} & 0 \\ \dot{\gamma} & 0 & 0 \\ 0 & 0 & 0 \end{bmatrix} \tag{4.7}$$

the shear stress is

$$\sigma(t) = \int_{-\infty}^{t} G(t-t')\dot{\gamma}(t') \, dt' \tag{4.8}$$

and for oscillatory shear is

$$\sigma(t) = \gamma_0\omega \int_{-\infty}^{t} G(t-t') \cos(\omega t') \, dt' \tag{4.9}$$

Neglecting the starting transient which affects the first few cycles, it can be shown that[1]

$$G'(\omega) = \omega \int_{0}^{\infty} G(s) \sin(\omega s) \, ds \tag{4.10a}$$

with the loss modulus

$$G''(\omega) = \omega \int_{0}^{\infty} G(s) \cos(\omega s) \, ds \tag{4.10b}$$

Hence, behavior in small-amplitude oscillatory shear can all be deduced from the relaxation modulus, $G(s)$.

4.1.4. Nonlinear Viscoelasticity

For some polymeric liquids, strain amplitude dependence is seen in both the amplitude ratio, $|G^*(\omega, \gamma_0)|$ and the phase delay, $\delta(\omega, \gamma_0)$ when γ_0 is not small. However, for most polymeric liquids, the stress response in large-amplitude oscillatory shear is not sinusoidal. This is why the property definitions for linear behavior, $|G^*|$ and δ, are not meaningful for a nonlinear viscoelastic response. A few cycles after starting the test, the shear stress normally becomes a standing wave which can be represented as a Fourier series of odd harmonics:

$$\sigma(t) = \sum_{m=1,\text{odd}}^{\infty} \sigma_m \sin(m\omega t + \delta_m) \tag{4.11}$$

where the amplitudes, $\sigma_m(\omega, \gamma_0)$, and phase contents, $\delta_m(\omega, \gamma_0)$, of the harmonics depend upon both strain amplitude and frequency. For liquids, only odd-valued harmonics are observed, so a plot of shear stress *versus* shear rate always gives a two-fold symmetric loop.

The cyclic integral of the shear stress with respect to the shear strain gives the lost work per cycle, W_L:

$$W_L = \oint \sigma \, d\gamma - \pi \sigma_1 \gamma_0 \sin(\delta_1) \tag{4.12}$$

Hence, all the lost work is in the first harmonic.

4.1.5. Normal Stress Differences

The Boltzmann superposition principle predicts that the first and second normal stress differences, N_1 and N_2, in any simple shear flow are zero, where

$$N_1 = \tau_{11} - \tau_{22} \tag{4.13a}$$

and

$$N_2 = \tau_{22} - \tau_{33} \tag{4.13b}$$

The Lodge rubber-like liquid theory predicts that in oscillatory shear, the shear stress is sinusoidal and given by eqn (4.3), while the first normal stress difference is nonzero and given by

$$N_1(t) = \gamma_0^2 [G'(\omega) - B(\omega) \cos(2\omega t) - C(\omega) \sin(2\omega t)] \tag{4.14}$$

where $B(\omega)$ and $C(\omega)$ are integral transforms of the memory function. Thus, the first normal stress difference oscillates at twice the frequency of the strain and has a nonzero offset or average value that is equal to $\gamma_0^2 G'(\omega)$. While the rubber-like liquid model predicts that the second normal stress difference is zero, this theory can be simply modified by replacing the Finger tensor by a linear combination of the Finger and Cauchy tensors with the result that

$$N_2(t) = \beta N_1(t) \tag{4.15}$$

where β is a constant. This model constitutes a second-order theory of nonlinear viscoelasticity and suggests that there may be a range of strain amplitudes in which the shear stress is still sinusoidal and given by eqn (4.3) but in which N_1 and N_2 have measurable magnitudes and are sinusoidal at a frequency twice that of the excitation.

The direct measurement of normal stress differences in transient flows is quite difficult, but if the fluid exhibits stress birefringence, an optical technique can be used to measure these quantities. In particular one makes use of the stress-optical law, given by

$$n_{ij} = C\tau_{ij} \qquad (4.16)$$

where n_{ij} is the birefringence in the x_i–x_j plane, and C is the stress-optic coefficient. The most easily measured component of the birefringence is the one proportional to $(\tau_{11} - \tau_{33})$, which is called N_3. According to the second-order theory of viscoelasticity mentioned above, this is equal to $N_1(1 + \beta)$. Kornfeld et al.[2] used this approach to measure $N_3(t)$ for a series of molten polymers. In addition, they proposed a procedure for determining $G'(\omega)$ and $G''(\omega)$ from this result.

For larger amplitudes, well into the nonlinear regime, we expect the normal stress differences to contain an increasing number of higher harmonics, at even multiples of ω.

4.2. EXPERIMENTAL ERRORS

4.2.1. Fluid Inertia

For simple shear in which fluid inertia cannot be neglected, the equation of motion is

$$\rho \frac{\partial u_1}{\partial t} = \frac{\partial \sigma_{12}}{\partial x_2} \qquad (4.17)$$

which shows that fluid inertia can interfere with the desired generation of a homogeneous deformation. To minimize fluid inertial effects, one need only make the gap sufficiently small. The periodic part of the solution to eqn (4.17) for linear viscoelastic behavior with one plate moving with velocity $U = U_0 \sin(\omega t)$ gives the shear rate profile[3]

$$\dot{\gamma}(t, x_2) = -U_0\sqrt{\alpha^2 + \beta^2} \left[\frac{\cosh(2\alpha x_2) + \cos(2\beta x_2)}{\cosh(2\alpha h) + \cos(2\beta h)} \right] \cos(\omega t + \Phi) \qquad (4.18a)$$

where

$$\Phi = -\Phi_1 + \Phi_2 + \Phi_3 \qquad (4.18b)$$

and

$$\Phi_1 = \arctan(\alpha/\beta) \tag{4.18c}$$

$$\Phi_2 = \arctan[\tan(\beta x_2)\tanh(\alpha x_2)] \tag{4.18d}$$

$$\Phi_3 = \arctan[\tanh(\alpha h)/\tan(\beta h)] \tag{4.18e}$$

$$\alpha = \omega\sqrt{\rho\gamma_0/\sigma_0}\,\sin(\delta/2) \tag{4.18f}$$

$$\beta = \omega\sqrt{\rho\gamma_0/\sigma_0}\,\cos(\delta/2) \tag{4.18g}$$

For linear viscoelastic behavior, β, the shear wavelength, L_s, and the shear wave velocity, v_s, are related by

$$\beta = 2\pi/L_s = \omega/v_s \tag{4.19}$$

Hence, the fractional deviation in shear-rate amplitude from homogeneous simple shear:

$$\varepsilon_i = \frac{\beta h\sqrt{[(\alpha/h)^2 + 1]}\,\sqrt{[\cosh(2\alpha x_2) + \cos(2\beta x_2)]}}{\sqrt{[\cosh(2\alpha h) - \cos(2\beta h)]}} \tag{4.20}$$

can be used to calculate just how small the gap must be to minimize inertial errors.

Important design results are : (1) for ε_i less than $0\cdot01$ and phase error less than one degree, L_s must be less than $20h$ and (2) a standing wave is obtained when t is greater than $4h/v_s$. Nomographs, helpful for determining $\varepsilon_i(x_2)$, are provided by Schrag.[3] Also, Darby[4] and Darby and Shah[5] gives explicit rheometer design constraints for some special cases of eqn (4.20).

Unfortunately, since there is no unifying theory for nonlinear viscoelasticity, there is no general design equation for minimizing inertial effects in large-amplitude oscillatory shear. Hence, the above criteria are valid only for linear behavior. However, we believe they are still useful guidelines when nonlinear behavior is exhibited. In particular, we propose that in the definitions of α and β, the loss angle be replaced by the angle δ_L defined by eqn (4.32). In any event, one can verify experimentally that there is no significant error due to inertia by carrying out experiments at a single γ_0 but with two gap spacings and comparing the stress waveforms. The large-amplitude oscillatory shear test results should be independent of the gap spacing.

4.2.2. Viscous Heating
When a liquid is deformed, its temperature rises. Since rheological properties are strong functions of temperature, the effects of viscous dissipation must be minimized when making rheological measurements.

Assuming specific forms for the frequency and temperature dependencies of $G'(\omega, T)$ and $G''(\omega, T)$, Schapery and Cantey[6] carried out a detailed analysis of viscous heating of a linear viscoelastic liquid in oscillatory shear. Defining a dimensionless quantity that is proportional to the square of the shear stress amplitude:

$$g \propto \sigma_0^2 \qquad (4.21a)$$

and a dimensionless quantity that is proportional to the square of the strain amplitude:

$$\Gamma \propto \gamma_0^2 \qquad (4.21b)$$

Schapery found:

$$g = [2/(\Gamma+1)]\{\ln[\sqrt{(\Gamma+1)} + \sqrt{\Gamma}]\}^2 \qquad (4.22)$$

Hence, for this special case, Schapery showed that viscous heating causes the shear-stress amplitude to go through a maximum with strain amplitude. When the strain amplitude exceeds this value, the sample temperature eventually increases without bound, a condition called thermal runaway. Under these conditions it might still be possible to make useful measurements before the onset of thermal runaway. Schapery showed that the time before the onset of thermal runaway is proportional to the square of the ratio of the gap spacing to the shear stress amplitude:

$$t \propto (h/\sigma_0)^2 \qquad (4.23)$$

Schapery's analysis neglects flow field heterogeneity caused by viscous dissipation. In fact, we expect the fluid near the adiabatic boundary to undergo larger deformations than the rest of the sample. This makes it particularly important to minimize the effects of viscous heating. Since the fluid at the midplane is hotter, we expect it to undergo larger deformations.

In practice, viscous dissipation effects can cause the observed value of $\sigma_1(\omega, \gamma_0)$ to decrease with the number of oscillations. Also, erroneously low values of $\sigma_1(\omega, \gamma_0)$ will be observed when the sample is too thick.

3. MEASUREMENT TECHNIQUES

The previous sections show how to avoid errors due to fluid inertia and viscous heating in large-amplitude oscillatory shear experiments. However, we are still presented with a formidable experimental challenge. No matter which geometry we use, the flow near the edges of the sample is impossible to control in any deformation incorporating large shear strains or large shear rates, e.g. large-amplitude oscillatory shear. When material properties are inferred from total force or total torque measurements, these edge effects cause significant errors which cannot be corrected.

Cone–plate flow has been used to measure large-amplitude oscillatory shear properties. For instance, MacSporran and Spiers[7] used a Weissenberg rheogoniometer to characterize polymer solutions in large-amplitude oscillatory shear with cone–plate flow. For low-viscosity liquids, centripetal acceleration causes sample outflow. For polymer melts normal stress effects and edge fracture distort the free boundary in large-amplitude oscillatory shear.[8] Hence, oscillatory-shear measurements on polymer melts are very difficult to make for strain amplitudes much above one using cone–plate flow. Hence, cone–plate flow is not a useful technique for polymer melts when $\gamma_0 > 1$.

Unlike cone–plate flow, however, parallel-disk flow is a heterogeneous flow field. The shear-strain amplitude is proportional to radial distance from the center. Since large-amplitude oscillatory shear properties are themselves strain-amplitude dependent, this greatly complicates the analysis. The equations required to extract nonsinusoidal stress responses to large amplitude oscillatory shear from the observed torque in parallel-disk flow, have been worked out by MacSporran and Spiers. We note that it is necessary to take the derivative of every component of the torque with respect to the strain amplitude at the rim. Thus, torque measurements with very high signal-to-noise levels, and several experiments, are required. For polymer solutions, strain amplitudes up to 27 are reported. Measurements at much lower strain amplitudes were corroborated with cone–plate flow data.

Less general schemes to extract nonsinusoidal large-amplitude oscillatory properties from parallel-disk flow experiments have been developed by Powell and Schwarz,[9,10] and by Spiers,[11] who assumed specific constitutive relations to relate the measured torque and thrust, to the shear stress and the first normal stress difference. Since a single constitutive equation will not be valid for all fluids, the more general approach of MacSporran and Spiers seems preferable for parallel-disk flow.

Some of the most successful attempts to measure large-amplitude oscillatory shear properties have employed concentric cylinder rheometers. For example, Onogi and coworkers[12,13,14] used concentric cylinders for nonsinusoidal oscillatory shear measurements on polymer solutions and suspensions for shear-strain amplitudes below unity. Dealy and coworkers[15,16] extended this method to molten plastics with a tedious sample-insertion technique, but they found that the Weissenberg effect causes a severe distortion of the free boundary when strains above 10 are used.

Sliding-cylinder flow has also been used for large-amplitude oscillatory shear but this flow is not homogeneous. McCarthy[17] used sliding-cylinder flow to measure nonsinusoidal stress responses for molten plastics in large-amplitude oscillatory shear, while Tsai and Soong[18] used this technique for polymer solutions. These researchers did not account for flow-field heterogeneity. Hibberd and coworkers[19,20] have shown that flow heterogeneity is not negligible for sliding-cylinder flow, and they developed a clever but tedious technique to correct for flow heterogeneity. Curiously, Hibberd and coworkers found that although highly strain-dependent, bread doughs do not exhibit nonsinusoidal stress responses in large-amplitude oscillatory shear. Thus, they could interpret their results in terms of γ_0-dependent storage and loss moduli.

Sliding-plate flow has also been used to measure large-amplitude oscillatory shear properties. As mentioned previously, in the absence of viscous heating and inertial effects, this flow generates homogeneous simple shear except near the edges of the sample. In most sliding-plate rheometers, the shear stress is inferred from a total force measurement and thus includes an error due to flow heterogeneity at sample edges. Another source of error is friction in the guide mechanisms. In order to avoid this error, a sandwich arrangement has been used where two samples are sheared between one moving plate and two stationary plates.[21] Liu *et al.*[22,23] used this type of apparatus to measure the large-amplitude oscillatory shear properties of polymer solutions with strain amplitudes up to 5. Sivashinsky *et al.*[24] used the same rheometer to measure stress growth following large-amplitude oscillatory shear on polymer solutions. Precision guide rods used to maintain plate parallelism caused errors in the total force measurement.

To avoid error due to flow heterogeneity near the edges, Dealy[25] invented a shear-stress transducer to be flush-mounted in the stationary plate. Hence, the shear stress can be measured locally in a region of uniform deformation, away from the free boundaries. This instrument has been used to measure the response of molten plastics to large-amplitude oscillatory shear.[26]

Giacomin and Dealy[27] and Giacomin[28] used this apparatus to compare the locally measured shear stress with the total force on the stationary plate and found large discrepancies for molten plastics in large-amplitude oscillatory shear. He attributed these discrepancies to flow-field hetero-

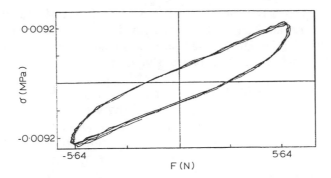

Fig. 4.2. Typical counterclockwise loop of shear stress *versus* total force. Polystyrene, 190°C, 0·2 Hz, $\gamma_0 = 7$.

geneity near the free boundaries. For instance, a sample of molten polystyrene with initial dimensions of 5·9 in × 3·4 in × 0·030 in was subjected to sinusoidal deformation with $\gamma_0 = 7$ and $\omega = 1·26$ rad/s at 190°C. Whereas the locally measured shear-stress amplitude was 9·20 kPa, the value inferred from the total force amplitude of 5·64 N was only 437 Pa. The large difference is attributed to the loss of effective contact area caused by the heterogeneous flow near the free boundaries. Figure 4.2 shows a plot of the shear stress, σ, *versus* the total force, F. The area of this loop is the experimental error due to the additional lost work caused by the heterogeneous deformation near the free boundaries.

4.4. METHODS OF DATA ANALYSIS

4.4.1. Spectral Analysis
Since the response to large-amplitude oscillatory shear depends upon both the strain and the frequency, the description of such a response is more complex than for the linear case, where $G'(\omega)$ and $G''(\omega)$ can be used. The most straightforward way to analyze large-amplitude oscillatory shear data is with spectral analysis. After a few oscillations, the stress response becomes a standing wave that is a Fourier series. For an isotropic liquid with fading memory, the series has only odd terms:

$$\sigma(t) = \sum_{m=1,\text{odd}}^{M} \sigma_m \sin(m\omega t + \delta_m) \tag{4.24}$$

While in theory M is infinity, in practice it is hard to resolve components where m is greater than 7. Hence one concise way of summarizing the shear-stress response to large-amplitude oscillatory shear is with the amplitudes, $\sigma_m(\omega, \gamma_0)$, and phase shifts, $\delta_m(\omega, \gamma_0)$, of the Fourier components.

The shear-stress wave is usually sampled digitally. Hence, the most convenient way to determine σ_m and δ_m is by use of a discrete Fourier transform. When σ is sampled N times with a constant interval, Δt, a time series, $\sigma(n\Delta t)$, is obtained, where n is the time sample index and its range is $0, 1, 2, \ldots, N-1$.

The discrete Fourier transform of $\sigma(n\Delta t)$ is

$$\sigma_d(k\Delta\omega) = \frac{1}{N} \sum_{n=0}^{N-1} \sigma(n\Delta t) \exp(-j.2\pi kn/N) \qquad (4.25)$$

where N is the number of samples, $j = \sqrt{-1}$, $\Delta\omega$ is the frequency resolution, $\Delta\omega = 2\pi/N\Delta t$, k is the discrete frequency component index and its range is $0, 1, 2, \ldots, N-1$. So the discrete Fourier transform, $\sigma_d(k\Delta\omega)$, is a set of complex numbers having the same dimensions as the stress, $\sigma(n\Delta t)$.

An interesting property of $\sigma_d(k\Delta\omega)$ is that its first component, $\sigma_d(0)$, is identically the arithmetic mean of the time series, $\sigma(n\Delta t)$. The highest, physically significant, frequency component of $\sigma_d(k\Delta\omega)$ occurs at $k = N/2$. This component, $\sigma_d(N\Delta\omega/2)$, is the right-most element of the discrete Fourier transform, and its frequency, $N\Delta\omega/2 = 1/2\Delta t$, is called the Nyquist frequency. When conducting large-amplitude oscillatory shear tests, the sampling interval must be low enough so that the Nyquist frequency exceeds the frequency of the highest harmonic of interest.

Since only the first $N/2$ points of $\sigma_d(k\Delta\omega)$ are physically meaningful, a time series containing N samples, $\sigma(n\Delta t)$, yields a discrete Fourier transform of $N/2$ complex numbers. Where the components of the discrete Fourier transform occur at multiples of the test frequency, ω, the material properties, σ_m, defined in eqn (4.24) can be inferred from the amplitudes of $\sigma_d(k\Delta\omega)$:

$$\sigma_m = 2 |\sigma_d(k\Delta\omega)| ; \qquad m = k\Delta\omega/\omega \qquad (4.26a)$$

Care must be taken to collect the time series so that the components of the discrete Fourier transform occur exactly at multiples of the test frequency, ω. This procedure is called frequency matching. The amplitude of each component of the discrete transform is

$$|\sigma_d(k\Delta\omega)| \equiv [\mathrm{Re}^2[\sigma_d(k\Delta\omega)] + \mathrm{Im}^2[\sigma_d(k\Delta\omega)]]^{1/2} \qquad (4.26b)$$

Obtaining the phase shifts of the higher harmonics, δ_m, defined in eqn (4.24) is a little trickier. Clearly, only the phase differences between the spectral components of stress and strain are material properties. A discrete Fourier transform must therefore also be applied to the shear strain to obtain its phase content, δ_γ. The phase contents of $\sigma_d(k\Delta\omega)$ are

$$\delta_d(k\Delta\omega) = \text{atan}\left[\frac{\text{Im}[\sigma_d(k\Delta\omega)]}{\text{Re}[\sigma_d(k\Delta\omega)]}\right] \tag{4.27}$$

where $m = k\Delta\omega/\omega$. The phase differences, δ_m, which are material properties are

$$\delta_m(\omega) = \delta_d(k\Delta\omega) - m\delta_\gamma \tag{4.28}$$

So, for the first three odd harmonics

$$\delta_1 = \delta_d(\omega) - \delta_\gamma \tag{4.29a}$$

$$\delta_3 = \delta_d(3\omega) - 3\delta_\gamma \tag{4.29b}$$

and

$$\delta_5 = \delta_d(5\omega) - 5\delta_\gamma \tag{4.29c}$$

This procedure is called frequency modulation. By convention, reported phase differences are between 0 and 2π.

Figure 4.3 shows an amplitude spectrum for the shear-stress response in small-amplitude oscillatory shear. Even in this case, the first, third, fifth, and seventh harmonics are detectable. Table 4.1 summarizes a study of the effect of γ_0 in the nonlinear range in large-amplitude oscillatory shear with the results of the discrete Fourier transform analysis. Figures 4.4 and 4.5 show the effect of γ_0 on $\sigma_m(\omega, \gamma_0)$ and $\delta_m(\omega, \gamma_0)$, respectively.

4.4.2. Error Analysis

Amplitude spectra are linearly additive. Hence, unwanted components of the amplitude spectrum, such as electrical interference appearing at multiples of 60 Hz, are easily identified and removed in the frequency domain. Since random noise has a uniform amplitude spectrum, it is also easily identified and removed in the frequency domain. Using the discrete Fourier transform in this way usually gives an improvement of two orders of magnitude in shear-stress resolution for large-amplitude oscillatory shear.

Additionally, the small phase errors due to the time delays of electronic amplifiers and transducers can also be easily eliminated in the frequency

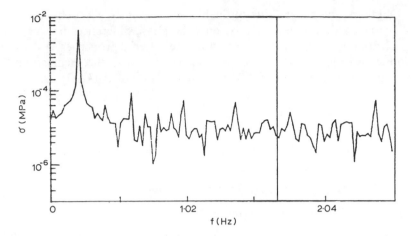

Fig. 4.3. Amplitude spectrum rich with higher harmonics for shear stress response to small-amplitude oscillatory shear for the molten polystyrene at 190°C and at a test frequency of 2 Hz with $\gamma_0 = 0.2$.

TABLE 4.1
Effect of γ_0 in Large Amplitude Oscillatory Shear
Polystyrene, 190°C, 0.2 Hz

γ_0	σ_1, MPa	σ_3, MPa	δ_1, rad	δ_3, rad
0.5	0.00225	0	0.9299	6.143
1.0	0.00451	0.000045	0.9522	1.238
2.5	0.00924	0.000222	1.072	2.081
4.0	0.01263	0.000561	1.158	2.572
5.0	0.01448	0.000855	1.252	2.990
6.0	0.01618	0.001088	1.256	2.870
7.0	0.01782	0.001380	1.200	3.335

domain. The dynamic response of electronic circuitry is usually described in terms of the -3 dB attenuation frequency, f_{3dB}. The time constant for each component is then $\tau_e = 1/f_{3dB}$, and the corresponding phase correction for each element of the discrete spectrum is

$$\delta_e = -\text{atan}(k\Delta\omega\tau_e) \qquad (4.30)$$

To compensate for attenuation, the amplitude spectrum must be multiplied by $\sqrt{(1 + k^2\Delta\omega^2\sigma_e^2)}$. Whereas attenuation is usually slight, eqn (4.30) shows that phase corrections for electronic circuitry can matter at surprisingly low frequencies.

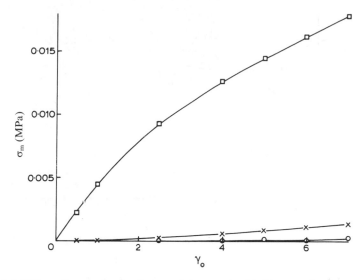

Fig. 4.4. Effect of γ_0 on the fundamental (squares), third harmonics (crosses) and fifth harmonics (circles) of shear stress in large-amplitude oscillatory shear for polystyrene at 190°C; $f = 0.2$ Hz.

4.4.3. Response Loops for Material Characterization

A frequently used graphical analysis technique is a plot of σ *versus* γ. Figure 4.6 shows some σ–γ loops selected from Table 4.1. Note that the nonlinearities are not clearly visible. A time trace of $\sigma(t)$ makes the nonlinearities even less clear. For liquids, a more useful technique is to plot σ *versus* $\dot{\gamma}$ as shown in Fig. 4.7. The nonlinearity of the response can be clearly seen as a deviation from an ellipse. Although the higher harmonics are not more than 8% of the fundamental, they profoundly affect the shape of the loop. Higher harmonics of comparable magnitude are sometimes dismissed as unimportant when only the trace is considered.

4.4.4. Analog Methods

Philippoff[29] used an analog computer to determine σ_m and δ_m by matching synthesized waveforms with observations. Analog circuits have also been used for real-time frequency analysis of nonlinear shear-stress waves of viscoelastic fluids in oscillatory shear.[30,31]

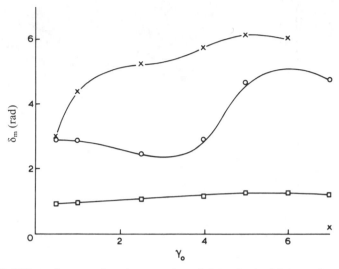

Fig. 4.5. Effect of γ_0 on the phase angles of the principal harmonics of shear stress in large-amplitude oscillatory shear for polystyrene at 190°C; $f = 0.2$ Hz. Fundamental (squares), 3rd harmonic (crosses), and 5th harmonic (circles).

4.4.5. Time-Domain Analysis

In principle, σ_m and δ_m can also be obtained by fitting the time series, $\sigma(n\Delta t)$, with a truncated Fourier series. However, since this requires nonlinear regression, solutions for σ_m and δ_m are not unique. This is considerably more complicated than using the discrete Fourier transform.

4.4.6. Approximate Methods

Instead of measuring all frequency components for the stress response, only certain gross features of the loop are sometimes reported. When determining the viscous dissipation in rubbers under vibratory loads, for example, one might only be interested in the lost work per cycle, the higher harmonics making no contribution to the lost work.[32,33] While the higher harmonics are an interesting feature of the nonlinear response for certain plastics, for simplicity, several approximate techniques for describing nonlinear behavior are in common use.

Sometimes only the stress amplitude, σ_0, and the lost work per cycle, W_L, or simple functions of these are reported.[34,17] The lost work per cycle is easily deduced from the area of the stress *versus* strain loop which can be obtained manually using a planimeter on σ–γ loops.[29] Using a simple

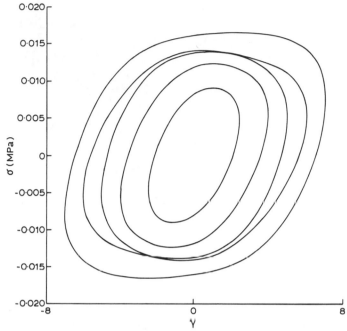

Fig. 4.6. Shear stress *versus* shear strain loops for molten polystyrene in oscillatory shear at 190°C. Strain amplitudes (from innermost to outermost loops) are 2·5, 4, 5, 6, and 7. Constant frequency of 0·2 Hz.

function of σ_0 and W_L, Thurston[35] and Thurston and Pope[36] pioneered the use of three-dimensional drawings to illustrate the large-amplitude oscillatory shear behavior of biofluids with ω and $\dot{\gamma}_0$ on separate axes. Stereoscopic pictures were used to provide the depth perception required to separate the effects of ω and $\dot{\gamma}_0$.

The most common way to describe nonlinear viscoelastic behavior in the literature is by use of storage and loss moduli defined in terms of an average loss angle:

$$G'(\omega, \gamma_0) \equiv (\sigma_0/\gamma_0)\cos\delta_L \qquad (4.31a)$$

and

$$G''(\omega, \gamma_0) \equiv (\sigma_0/\gamma_0)\sin\delta_L \qquad (4.31b)$$

where

$$\delta_L \equiv \mathrm{asin}(W_L/\pi\sigma_0\gamma_0) \qquad (4.32)$$

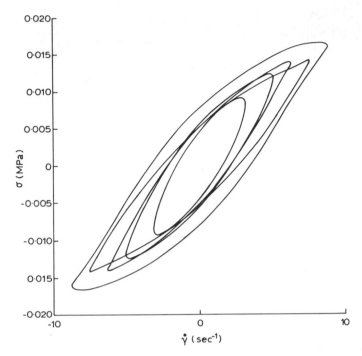

Fig. 4.7. Shear stress *versus* shear rate loops for molten polystyrene in oscillatory shear at 190°C. Strain amplitudes (from innermost to outermost loops) are 2·5, 4, 5, 6, and 7. Constant frequency of 0·2 Hz.

The problem with approximate methods is that they give no information about the shape of the stress response, an integral part of the nonlinear viscoelasticity.

Another approach is to report the root-mean-square value, σ_{rms}, of $\sigma(t)$ as a function of ω and γ_0. This is a common way of analyzing waveforms in electrical engineering, and appropriate analog circuitry is readily available to calculate σ_{rms} directly.[37]

4.5. PLAUSIBLE PHASE ANGLES

The second law of thermodynamics requires that $W_L \geq 0$. Combining this inequality with (4.12) gives:

$$\sin \delta_1 \geq 0 \tag{4.33}$$

Hence

$$0 \leq \delta_1 \leq \pi \qquad (4.34)$$

In practice, measured values of the phase angle for the first harmonic are always in the first quadrant:

$$0 \leq \delta_1 \leq \pi/2 \qquad (4.35)$$

and most constitutive equations predict this too.

4.6. THE PIPKIN DIAGRAM

It is useful to construct a map showing the various regimes of behavior that can be seen in large-amplitude oscillatory shear. Pipkin has suggested a plot of $\dot{\gamma}_0$ *versus* ω as a basis for such a map, as shown in Fig. 4.8. On this diagram, γ_0 is constant along straight lines through the origin.

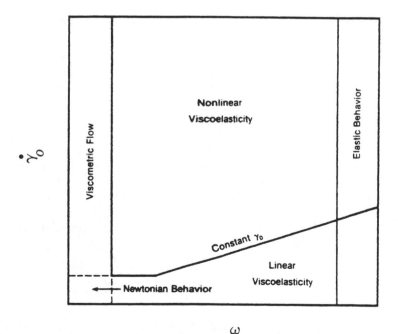

Fig. 4.8. Pipkin diagram showing regimes of behavior for oscillatory shear. (Not to scale; viscometric flow and linear viscoelasticity zones are actually quite narrow.)

At very low frequency, the flow is viscometric. If $\dot{\gamma}_0$ is also small, Newtonian behavior is seen. As frequency increases, some viscoelasticity appears. At low γ_0 we expect linear behavior. At some γ_0, we expect higher harmonics to arise.

Just to the right of the viscometric flow region, some authors claim that the stress is sinusoidal but that the storage and loss moduli depend upon strain amplitude, i.e. $G'(\omega, \gamma_0)$ and $G''(\omega, \gamma_0)$. Since no theory predicts such a regime, it would appear that the higher harmonics were simply overlooked. At very high frequencies we expect a purely elastic (non-dissipative) response. Increasing γ_0 in the nondissipative region causes nonlinear elasticity.

It has been suggested that for polymer melts at some critical stress amplitude, the melt will slip at the wall causing the stress signal to become erratic in oscillatory shear.[38] When this happens, it may be difficult to distinguish the effects of slip and nonlinear viscoelasticity.

4.7. INTERPRETING NONLINEAR BEHAVIOR

For a fundamental interpretation of rheological behavior, one requires a constitutive equation. Although a few constitutive equations have been used to predict large-amplitude oscillatory shear, none have been compared with experiments using strain amplitudes much higher than one. These predictions are frequently given in terms of $G'(\omega, \gamma_0)$ and $G''(\omega, \gamma_0)$ which have no unique definition for nonlinear behavior.

It is useful to have a general representation of the stress response in large-amplitude oscillatory shear. Separating each Fourier component into in-phase and out-of-phase parts, and factoring out γ_0, defines a set of nonlinear viscoelastic moduli:

$$\sigma(t) = \gamma_0 \sum_{m=1,\text{odd}}^{M} [G'_m(\omega, \gamma_0) \sin(m\omega t) + G''_m(\omega, \gamma_0) \cos(m\omega t)] \quad (4.36)$$

According to a number of constitutive theories, $G'_m(\omega, \gamma_0)$ and $G''_m(\omega, \gamma_0)$ can be expanded in odd powers of γ_0, and $\sigma(t)$ can thus be represented as:

$$\sigma(t) = \sum_{m=1,\text{odd}}^{M} \sum_{n=1,\text{odd}}^{m} \gamma_0^m [G'_{mn}(\omega) \sin(n\omega t) G''_{mn}(\omega) \cos(n\omega t)] \quad (4.37)$$

which nicely separates the strain dependence from the frequency dependence. However, the terms of this power series are not simply related to those of the Fourier series expressed by eqn (4.24). For instance, σ_1 and δ_1 depend upon the series of terms G'_{n1} and G''_{n1}.

The Doi–Edwards theory has been used to predict large-amplitude oscillatory shear behavior. Analyzing their experimental data using strain-dependent storage and loss moduli, Pearson and Rochefort[8] report good agreement for monodisperse polystyrene melts with γ_0 less than $1 \cdot 5$. The Doi–Edwards theory is of special interest because there are only two model parameters, and these can be determined from linear viscoelastic experiments. Predictions are reported in terms of explicit expressions for the functions defined in eqn (4.37), both with and without the use of the independent alignment approximation. Specifically, Helfand and Pearson[39] showed the independent alignment approximation is not valid for calculating G''_{31} and G'_{33} at high frequencies.

Yen and McIntire[40] used a BKZ model to calculate the response to large amplitude oscillatory shear using a five-parameter time-dependent elastic energy potential function. The parameters are determined from the linear spectrum and the viscometric functions $\eta(\dot{\gamma})$ and $N_1(\dot{\gamma})$, so the model contains no parameters that must be determined by fitting large-amplitude oscillatory shear data. Analyzing experimental data using strain-dependent storage and loss moduli, good agreement was reported for a few polymer solutions.

Several attempts have been made to explain the large-amplitude oscillatory shear behavior of melts and solutions by use of kinetic network theories. Lodge's rubber-like liquid model, an early kinetic network theory, does not predict nonlinearity in the shear-stress response in large-amplitude oscillatory shear. However, the model does predict a first normal stress difference in large-amplitude oscillatory shear which is given by eqn (4.14). Hence, this model requires only the linear spectrum.

Acierno *et al.*[41] proposed a kinetic network theory which requires the linear spectrum and the viscosity curve to fit one nonlinear parameter. Tsang and Dealy[42] evaluated this model for a polystyrene melt in large-amplitude oscillatory shear. Since they used σ–$\dot{\gamma}$ loops for both their data and the model evaluation, a comparison of the shapes of $\sigma(t)$ was possible. Good agreement was reported for the predicted loop shapes but the overall stress amplitudes do not agree with experiment for large strains, $5 < \gamma_0 < 10$. Specifically, the model underpredicts the lost work per cycle. Interestingly, the large-amplitude oscillatory shear test brought out a deficiency in this model which no other test brought out. Thus, the large-amplitude oscillatory shear test is considered a particularly discriminatory test for constitutive theories.

Several alternative kinetic rate expressions have been used to predict large-amplitude oscillatory shear. Liu *et al.*[23] developed one for simple shear flows which requires the linear spectrum, and three nonlinear parameters. A nonlinear oscillatory shear test, and two viscometric functions are required to determine the three parameters. Predicted $\sigma-\gamma$ loops agreed well with measured ones for polymer solutions at $\gamma_0 = 3$.

Another kinetic network theory, the Phan–Thien Tanner model,[43] was also evaluated by Tsang and Dealy.[42] While poor agreement is reported for the loop shapes, the stress amplitude predictions were excellent.

A few well-known integral forms of the kinetic network theory have also been evaluated in large-amplitude oscillatory shear. Analyzing experimental data using strain-dependent storage and loss moduli, poor agreement was reported for polymer solutions with $\gamma_0 \leq 1.5$ for the Bird–Carreau,[34] Carreau-B,[44] and MacDonald–Bird–Carreau[45] models. Fan and Bird[46] evaluated the Curtiss–Bird constitutive equation in large-amplitude oscillatory shear, but this has yet to be compared with experimental data.

REFERENCES

1. J. D. Ferry, *Viscoelastic Properties of Polymers*, 3rd Ed., John Wiley & Sons, New York, 1980.
2. J. A. Kornfeld, G. G. Fuller and D. S. Pearson, *Rheol. Acta*, 1990, **29**, 105.
3. J. L. Schrag, *Trans. Soc. Rheol.*, 1977, **21**(3), 399.
4. R. Darby, *Viscoelastic Fluids: An Introduction to Their Properties and Behavior*, Marcel Dekker, Inc., New York, 1976, p. 251.
5. B. H. Shah and R. Darby, *Polym. Eng. Sci.*, January 1976, **16**(1), 46.
6. R. A. Schapery and D. E. Cantey, *AIAA J.*, February 1966, **4**(2), 255–64.
7. W. C. MacSporran and R. P. Spiers, *Rheol. Acta*, 1984, **23**, 90.
8. D. S. Pearson and W. E. Rochefort, *J. Polym. Sci.: Pol. Phys. Ed.*, 1982, **20**, 83.
9. R. L. Powell and W. H. Schwarz, *J. Polym. Sci.: Pol. Phys. Ed.*, 1979, **17**, 969.
10. R. L. Powell and W. H. Schwarz, *J. Rheol.*, 1979, **23**(3), 323.
11. R. P. Spiers, PhD Thesis, University of Bradford, 1977.
12. S. Onogi, T. Masuda and T. Matsumoto, *Trans. Soc. Rheol.*, 1970, **14**(2), 275.
13. T. Matsumoto, Y. Segawa, Y. Warashina and S. Onogi, *Trans. Soc. Rheol.*, 1973, **17**(1), 47.
14. S. Onogi and T. Matsumoto, *Polym. Eng. Revs*, 1981, **I**(1), 45.
15. T.-T. Tee and J. M. Dealy, *Trans. Soc. Rheol.*, 1975, **19**(4), 595.
16. J. M. Dealy, J. F. Petersen and T.-T. Tee, *Rheol. Acta*, 1973, **12**, 550.
17. R. V. McCarthy, *J. Rheol.*, 1978, **22**(6), 623.

18. A. T. Tsai and D. S. Soong, *J. Rheol.*, 1985, **29**(1), 1.
19. G. E. Hibberd, W. J. Wallace and K. A. Wyatt, *J. Sci. Instrum.*, February 1966, **43**, 84.
20. G. E. Hibberd and N. S. Parker, *Cereal Chem.*, 1975, **52**(3-II), 1r.
21. J. M. Dealy and A. J. Giacomin, *Rheological Measurement*, Ch. 12, A. A. Collyer and D. W. Clegg (eds), Elsevier Applied Science, London, 1988.
22. T. Y. Liu, D. W. Mead, D. S. Soong and M. C. Williams, *Rheol. Acta*, 1983, **22**, 81.
23. T. Y. Liu, D. S. Soong and M. C. Williams, *J. Polym. Sci.: Pol. Phys. Ed.*, 1984, **2**, 1561.
24. N. Sivashinsky, A. T. Tsai, T. J. Moon and D. S. Soong, *J. Rheol.*, 1984, **28**(3), 287.
25. J. M. Dealy, US Patent No. 4 464 928, August 14, 1984.
26. A. J. Giacomin, T. Samurkas and J. M. Dealy, *Polym. Eng. Sci.*, April 1989, **29**, 499.
27. A. J. Giacomin and J. M. Dealy, Paper G3, 58th Annual Meeting Soc. Rheol. Tulsa, OK, USA, October, 1986.
28. A. J. Giacomin, *A Sliding Plate Melt Rheometer Incorporating a Shear Stress Transducer*, Doctoral Dissertation, Dept. of Chemical Engineering, McGill University, Montreal, Canada, June 1987.
29. W. Philippoff, *Trans. Soc. Rheol.*, 1966, **10**(1), 317.
30. K. Walters and T. E. R. Jones, Proc. 5th Int. Cong. Rheol., Vol. 4, S. Ohogi (ed), University of Tokyo Press, Tokyo, 1970, p. 337.
31. J. Harris and K. Bogie, *Rheol. Acta*, 1967, **6**(1), 3.
32. P. J. Cain, Bull. No. 300014-56 170.70-02, MTS Systems Corp., Eden Prairie, Minnesota, 1986.
33. D. O. Stalnaker and T. S. Fleischman, *Closed Loop*, **14**(2), MTS Systems Corp., Eden Prairie, Minnesota, 1985, p. 4.
34. I. F. MacDonald, B. D. Marsh and E. Ashare, *Chem. Eng. Sci.*, 1969, **24**, 1615.
35. G. B. Thurston, *J. Non-Newtonian Fluid Mech.*, 1981, **9**, 57.
36. G. B. Thurston and G. A. Pope, *J. Non-Newtonian Fluid Mech.*, 1981, **9**, 69.
37. R. Heinrich, pers. comm.
38. S. G. Hatzikiriakos and J. M. Dealy, *J. Rheol.*, 1991, **35**(4), 497.
39. E. Helfand and D. S. Pearson, *J. Polym. Sci.: Polym. Phys. Ed.*, 1982, **20**, 1249.
40. H.-C. Yen and L. V. McIntire, *Trans. Soc. Rheol.*, 1972, **16**(4), 711.
41. D. Acierno, F. P. La Mantia, G. Marrucci and G. Titomanlio, *J. Non-Newtonian Fluid Mech.*, 1976, **1**, 147.
42. W. K.-W. Tsang and J. M. Dealy, *J. Non-Newtonian Fluid Mech.*, 1981, **9**, 203.
43. N. Phan-Thien, *J. Rheol.*, 1978, **22**, 259.
44. I. F. MacDonald, *Rheol. Acta*, 1975, **14**, 801.
45. I. F. MacDonald, *Rheol. Acta*, 1975, **14**, 906.
46. X.-J. Fan and R. B. Bird, *J. Non-Newtonian Fluid Mech.*, 1984, **15**, 341.

Chapter 5

A Parsimonious Model for Viscoelastic Liquids and Solids

H. H. Winter, M. Baumgärtel

*Department of Chemical Engineering, University of Massachusetts,
Amherst, Massachusetts, USA*

AND

P. R. Soskey

*EniChem Americas Inc., 2000 Princeton Park Corporate Center,
Monmouth Junction, New Jersey, USA*

5.1. INTRODUCTION

The enormous progress in the capabilities of commercial rheometers allows measurements of rheological material functions with a high degree of reproducibility and over wide ranges of time and temperature. However, rheometry is still very tedious and its usefulness is often questioned. It seems that nearly all efforts had concentrated on the advancement of rheometers and too little thought had been given to questions like 'how should we analyze the data?', 'how can we combine or compare rheometrical experiments?' or 'how can we make better use of the data?'. The problem is not that methods for the evaluation and application of the rheological data are insufficiently understood, at least in the linear viscoelastic range. The problem is that these methods still require too many repetitive and time-consuming steps to be performed on a routine basis. For example, the determination of relaxation and retardation time spectra was mostly seen in connection with major theses.

Here we assign the repetitive part of data processing to a self-contained computer program and, thus, become liberated from the non-creative chores. All energy and enthusiasm should go towards the actual evaluation and application of the data. This can be facilitated by the interactive graphics tools of state-of-the-art personal computers. The proposed computer program is in use in over 40 laboratories.

The main objective of this chapter is to describe the rheological behavior with as few parameters as possible and to determine these parameters. Such a *parsimonious model* will be attempted by means of computer-aided methods which are easy to use by a technician in the rheometry laboratory or by a researcher who wants to explore already existing rheological material data. In a sequence of well-defined steps, one begins by transferring rheometer data, independently of the rheometer of origin, into a standard format. As an alternative, one is allowed to prepare one's own input file from archived data. Then, data can be time–temperature superpositioned for determination of the activation energy E_a/R or the WLF coefficients. Depending on the type of rheological data, one proceeds to evaluate the power-law parameters from steady shear viscosity or the discrete relaxation time spectrum from the dynamic moduli G', $G''(\omega)$. The discrete spectrum allows prediction of linear stress responses in start-up of shear followed by relaxation, creep followed by recovery, and step strain, to name a few examples.

Section 5.2 of this chapter defines the most commonly used constitutive relations for steady shear, linear viscoelasticity, and finite viscoelasticity of polymer melts and concentrated solutions. The selected equations were chosen since they are often used in model calculations. The parameters in these equations seem to be most urgently needed. Section 5.3 of the chapter guides through the determination of these material parameters, aided by the computer. Section 5.34 gives a few applications of the rheological data, heavily using computer graphics in an interactive mode. The appendix contains definitions of the most important vector and tensor variables.

5.2. PARAMETERS IN RHEOLOGICAL MODELS

5.2.1. Modelling of the Stress

Computer simulations of polymer processing often attempt to calculate the stress distribution and the motion of processing flows.[1] The stress in a deforming polymer element depends upon the rate of deformation and on the molecular mobility along the path line of this material element in its flow. The rate of deformation is prescribed by the specific flow conditions and by conservation of mass and momentum. The mobility of macromolecules depends upon their architecture, on the degree of order, and on the distance from the glass transition temperature. It is especially this sensitivity to molecular parameters which magnifies small differences between polymers during their processing.

The relation between stress, strain history, and temperature history is called a rheological model or, more formally, a rheological constitutive equation. Unfortunately, there does not seem to exist a constitutive equation for polymer melts or solutions which is universally valid. Many different ones have been proposed, each of them having their advantages and their shortcomings. The basic rules which constitutive equations have to satisfy, their underlying physics, and their predictive power have been discussed in great detail by Bird *et al.*[2] and by Larson,[3] and the reader is referred to these sources for deeper study. Computer simulations require a rheological model which most easily complies with numerical codes. One prefers the most simple rheological model which still captures the basic features of the process of interest. Therefore, special emphasis will be given to the 'user-friendliness' of individual models. The model parameters should be easy to determine and they should lend themselves to model calculations.

5.2.1.1. Three Types of Rheological Limiting Behavior

There are few ideal constitutive equations. These are worth mentioning first since they give a framework for the entire field of rheology as shown in Fig. 5.1. The deviatoric stress, $\boldsymbol{\sigma}$, as function of strain rate, $\mathbf{D}$, or strain, $\mathbf{C}^{-1}$, is known for the limiting cases of

—the Newtonian liquid

$$\boldsymbol{\sigma}(t) = 2\eta \mathbf{D}(t) \tag{5.1}$$

—the Hookean solid (t_0 denoting the stress-free state)

$$\boldsymbol{\sigma}(t) = G_e \mathbf{C}^{-1}(t_0, t) \tag{5.2}$$

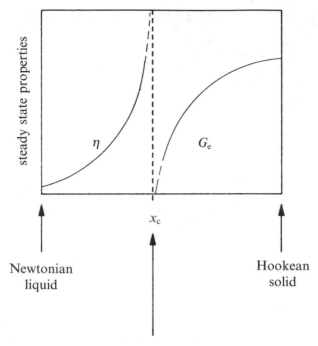

Network at the gel point

Fig. 5.1. Limiting rheological behavior of viscoelastic materials as shown with a crosslinking polymer (parameter x = degree of crosslinking). Representative properties are the steady shear viscosity for the liquid state (sol) and the equilibrium modulus for the solid state (gel). All viscoelastic liquid states are in between the Newtonian liquid ($x=0$) and the critical gel ($x=x_c$). Equivalently, all viscoelastic solids are in between the critical gel and the Hookean solid ($x=1$). The rheology at each of the three limiting states is universal and of extreme simplicity.

—the critical gel as transition state between liquid and solid[4,5]

$$\sigma(t) = S \int_{-\infty}^{t} (t-t')^{-n} 2\mathbf{D}(t') \, dt' \tag{5.3}$$

Material properties are represented by the viscosity, η, the modulus, G_e, the gel strength, S, and the relaxation exponent, n.

It should be noted that no material behaves exactly like any of these limiting cases: a low molecular weight liquid is mostly viscous but it also exhibits elasticity at very short time scales ($<10^{-10}$ s, for instance); an elastomer is mostly elastic but it does not store all its strain energy, and no polymer exists exactly at the gel point. Polymeric *liquids* behave somewhere in between the Newtonian liquid and the critical gel; polymer *solids* behave somewhere in between the critical gel and the Hookean solid. The intermediate behaviors are vastly more complex than the limiting states. Some of them will be discussed in the following in order to determine their material parameters.

5.2.1.2. Temperature Dependence
The properties in eqns (5.1)–(5.3) depend upon temperature, especially the viscosity which decays strongly with temperature (see Section 5.3.2) while the modulus might be approximated by a constant over a fairly wide temperature range. It is interesting to note that the stresses in the Newtonian liquid and in the Hookean solid depend upon the *instantaneous temperature* while the stress in the gel depends upon the *temperature history*. This history dependence is typical for viscoelastic materials.

5.2.1.3. Pressure Dependence and Deviatoric Stress σ
The effect of pressure is commonly neglected due to the small compressibility of polymers, and constitutive equations describe the stress only within an additive isotropic contribution

$$\tau = -p\mathbf{l} + \sigma \tag{5.4}$$

This simplification is acceptable for extrusion simulations; however, it will lead to errors when modelling the injection-molding process. In spite of this, constant density will be assumed in the following.

5.2.1.4. Polymers with Complex Superstructure
Most polymeric liquids adopt an equilibrium state at rest and their rheological behavior can be related to this equilibrium state as discussed in Section 5.3. However, there are many technically important polymers whose equilibrium state is extremely difficult to reach. In real experiments or during processing they adopt metastable states which depend upon their processing history. These metastable states determine the rheology and, when frozen in, the solid-state properties. Examples of such polymers are block copolymers, liquid crystalline polymers, polymer blends, and filled polymers, to name but a few. Their rheological behavior is most complex and their understanding is in its infancy. It will not be discussed

here. However, the techniques as discussed in this study are currently applied to such structured materials.[6]

5.2.2. Steady Shear Viscosity

The flow properties of polymers are advantageous for processing. At high rates of strain, such as during mixing inside an extruder or during filling of an injection mould, the viscosity is low and viscous dissipation is maintained at an acceptable level. On the other hand, the viscosity is very high at low stress (and hence low rate of strain). This high viscosity value is helpful, for example, in extrusion where the polymer extrudate, after having emerged from the shaping die, can support itself against gravity while it is slowly cooled to the solid state.

It has always been the objective of processing simulations to capture this phenomenon of increased molecular mobility at high rates of strain. The phenomenon is called 'shear thinning' since it occurs most drastically in shear flow and since it can be measured most easily in shear-flow geometries. The flow rates and the forces in polymer processing have been most successfully modelled with a shear thinning viscosity.[7,8] For that purpose, the viscosity in the Newtonian fluid equation was simply replaced by a viscosity function which depends upon shear rate $\dot{\gamma}$ or, in general, on the second invariant of the rate-of-deformation tensor, $\dot{\gamma} = \sqrt{2\mathbf{D}:\mathbf{D}}$. This leads to the *generalized Newtonian liquid*

$$\boldsymbol{\sigma}(t) = \eta(\dot{\gamma}, T)2\mathbf{D}(t) \qquad (5.5)$$

The viscosity depends strongly upon shear rate, temperature, and somewhat on pressure. We do not have data for discussing the pressure effect, which is small at extrusion processes, but might be important for injection moulding.

For most polymers, the temperature effect is very strong. It can be modelled phenomenologically with the temperature shift factors, a_T and b_T, which will be defined in sections 5.2.32 and 5.3.2. The model is based on the observation that viscosity curves, $\eta(\dot{\gamma}, T_1)$, $\eta(\dot{\gamma}, T_2)$, $\eta(\dot{\gamma}, T_3)$, ... which were measured at a set of constant temperatures T_1, T_2, T_3, ... all have the same shape. Therefore, a known viscosity curve at reference temperature T_0 may be shifted to a desired temperature level T by applying the following relation:[9]

$$\eta(\dot{\gamma}, T) = \eta(a_T\dot{\gamma}, T_0)a_T(T, T_0)/b_T(T, T_0) \approx \eta(a_T\dot{\gamma}, T_0)a_T(T, T_0) \quad (5.6)$$

The above relation is often simplified by neglecting the factor $b_T = \rho_0 T_0/(\rho T)$ which is close to unity. This uniform T-shift seems to be a good model for conventional homopolymers. Obviously, it is not valid over a temperature range in which a phase transition occurs. One has to be especially careful with block copolymers, blends, liquid crystalline polymers, ionomers, etc. The pressure dependence may be included in the shift factor; however, this is not of interest here.

The generalized Newtonian liquid is suited for calculating the shear stress in steady shear flows, i.e. for flows in which the polymer is subjected to a constant shear rate along its path (for a detailed definition see Bird *et al.*[2]). The steady shear viscosity is defined as the ratio of shear stress to shear rate. It has to be emphasized, however, that the generalized Newtonian liquid does *not* realistically predict normal stresses in shear nor does it realistically predict the stress in other types of flow, such as extensional flow. This limitation is well known and does not cause a problem as long as eqn (5.5) is used for modelling the shear stress in flows which closely resemble steady shear. Shear is one of the most widely used flow geometries in polymer processing. Examples are flow through pipes, slits, annuli (channels of constant cross-section) and drag flow between parallel surfaces (wire coating, Couette flow, etc.). Even if a processing flow is not a steady shear flow in its pure sense, some deviation from steady shear flow seems to be acceptable in the calculations. The remainder of this section (5.2.2) will be restricted to the discussion of the shear stress, τ.

A typical shear viscosity of a polymer melt is shown in Fig. 5.2. The constant value of the viscosity at low rates is called the *zero-shear viscosity*, η_0.

Several functions have been proposed for fitting the shear thinning region at high rates. The most simple one is the 'power law'. In many simulations, the finite viscosity at low rates is unimportant and the shear stress is simply modelled by

$$\tau = m\dot{\gamma}^n \qquad (5.7)$$

The power-law exponent of common polymers ranges from 0·2 to 0·6, depending upon the type of polymer, the molecular weight, the molecular weight distribution, etc.

This power-law relation facilitates explicit models of many processing flows and, hence, is a most popular engineering tool. However, it predicts an infinite viscosity in the limit of zero shear rate and therefore overpredicts the stress. The degree of overprediction depends upon the specific processing flow. It can easily be avoided in computer simulations by truncating the power law as it exceeds the zero-shear viscosity. Also,

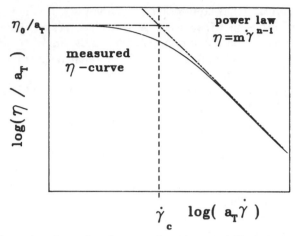

Fig. 5.2. Schematic of truncated power law for modelling the steady shear viscosity.

temperature dependence can be accounted for in a simple fashion. The viscosity for this type of non-isothermal modelling is sketched in Fig. 5.2 and is mathematically modelled by the *truncated power law*:

$$\eta(\dot{\gamma}, T) = \begin{cases} \eta_0(T_0)a_T & \text{for } (a_T\dot{\gamma}) < \dot{\gamma}_c & (5.8) \\ \eta_0(T_0)[\dot{\gamma}/\dot{\gamma}_c]^{n-1}a_T^n & \text{for } \dot{\gamma}_c < (a_T\dot{\gamma}) & (5.9) \end{cases}$$

The above equations simplify for the isothermal case with $a_T = 1$. The characteristic shear rate $\dot{\gamma}_c$ marks the onset of shear thinning.

The transition between zero-shear viscosity and shear thinning is more gradually fitted by the *Carreau model*.[10] Here it is modified to also describe the T-dependence:

$$\eta(\dot{\gamma}, T) = \eta_0(T_0)a_T[1 + (a_T\dot{\gamma}/\dot{\gamma}_c)^2]^{(n-1)/2} \qquad (5.10)$$

It is easy to check that the Carreau power law recovers eqn (5.8) at low rates and eqn (5.9) at high rates.

The parameters of the various power-law models are easily determined by numerical regression on the computer.

5.2.2.1. Characteristic Shear Rate
The characteristic shear rate is defined by the transition from η_0 to the power-law region as sketched in Fig. 5.2. It is of the same order of

magnitude as the reciprocal of the longest relaxation time (see eqn (5.15) below.

$$\dot{\gamma}_c \approx 1/\lambda_{max} \tag{5.11}$$

At low shear rates, $(\dot{\gamma}\lambda_{max}) \ll 1$, the polymer relaxes fast enough to remain in its equilibrium state, even during flow. At high rates, $1 < (\dot{\gamma}\lambda_{max})$, the molecular conformation is altered by the flow. The characteristic shear rate decreases with molecular weight ($\sim M^{3.4}$ for linear macromolecules) and strongly depends upon molecular structure.

The characteristic shear rate is strongly temperature-dependent.

$$\dot{\gamma}_c = \dot{\gamma}_c(T_0) \tag{5.12}$$

In the above equations, eqns (5.9) and (5.10), its value at the reference temperature T_0 was used.

5.2.2.2. Determination of the Steady Shear Viscosity
η is commonly measured with capillary rheometers, utilizing pressure flow through pipes or slits, and with rotational rheometers, using various geometries such as concentric cylinders, cone-and-plate, or concentric parallel disks. Capillary rheometry covers the high shear-rate and rotational rheometers cover the low shear-rate range. There are also empirical formulae which allow, with surprising accuracy, the estimation of the steady shear viscosity from the relaxation time spectrum. We will return to this point after having determined the spectrum.

5.2.3. Linear Viscoelasticity
The most basic information needed for modelling the elasticity of polymeric liquids provides of the relaxation time spectrum. We, therefore, start out by defining the relaxation time spectrum, explaining how it can be determined for a given polymer, and then showing the spectrum of several typical polymers.

The molecular mobility expresses itself in the relaxation time spectrum. It is well defined as long as perturbations from the equilibrium state are small, i.e. in the range of small deformations or for very slow flows. In this *linear viscoelastic* range, the constitutive equation is based on Boltzmann's superposition principle:[9]

$$\sigma(t) = \int_{-\infty}^{t} G(t-t')2\mathbf{D}(t') \, dt' \tag{5.13}$$

where σ is the stress tensor, $\mathbf{D}$, is the strain rate tensor, and $G(t - t')$ is the linear relaxation modulus. Using a discrete spectrum with N relaxation modes, the relaxation modulus $G(t - t')$ can be expressed as

$$G(t - t') = \sum_{i=1}^{N} g_i \exp[-(t - t')/\lambda_i] \qquad (5.14)$$

where coefficients, g_i, and relaxation times, λ_i, are the parameters of the discrete relaxation time spectrum. Here we chose exponential decays (Maxwell modes) since they approximate the data extremely well and they are simple to use in our computer program. The material is a solid if one of the maxwell models has an infinitely large relaxation time.

5.2.3.1. Deborah Number

The leading relaxation time, $\lambda_N = \lambda_{max}$, is often compared to the characteristic processing time t_p, such as the residence time of the polymer in a processing equipment or the inverse of the characteristic rate of deformation of the process. The ratio of these times is called the Deborah number

$$De = \lambda_{max}/t_p \qquad (5.15)$$

For processes with a small Deborah number, $De \ll 1$, the elasticity of the polymer is without consequence and the polymer might be modelled as a purely viscous liquid. For processes with a Deborah number of order one or above, elasticity cannot be neglected and a more complex constitutive equation has to be chosen for the flow simulation.

5.2.3.2. Temperature Dependence

The molecular mobility of amorphous polymers increases strongly with distance from the glass transition temperature, T_0. For most homopolymers, the entire relaxation spectrum shifts with the same temperature shift factor, a_T. Near the glass transition temperature, the temperature shift follows the WLF-Vogel relation[9]

$$\log a_T = -C_1(T - T_g)/(C_2 + T - T_g), \qquad T_g < T < (T_g + 100 \text{ K}) \qquad (5.16)$$

where $C_1 = 7.60$ and $C_2 = 227.3$ K for a wide range of polymers. At higher temperatures, an Arrhenius equation is sufficient:

$$\ln a_T = (1/T - 1/T_0)E_a/R, \qquad (T_g + 100 \text{ K}) < T \qquad (5.17)$$

Temperature-induced density changes result in a second shift factor

$$b_T = \rho_0 T_0/(\rho T) \qquad (5.18)$$

which is close to unity, in most cases. Ideally, this density change should be measured in a separate experiment by means of a densitometer. Such an instrument is not available in most laboratories and b_T has to be determined directly from the rheological data. An elegant method of determining b_T from G', $G''(\omega)$ data will be discussed in Section 5.3.2.

The temperature-dependent relaxation time spectrum is then written as

$$\lambda_i(T) = \lambda_i(T_0)a_T \quad \text{and} \quad g_i(T) = g_i(T_0)/b_T \qquad (5.19)$$

The magnitude of the shift factor is conventionally determined by shifting dynamic mechanical data which were measured at a set of different temperature levels. The frequency window of the data increases through this procedure, and the time window of the known relaxation time spectrum is extended accordingly.

5.2.3.3. Range of Applicability

The relaxation modes of the equilibrium state not only apply to the linear viscoelastic range but also describe transient behavior at large strain, since they depend upon the friction coefficient of the molecular chain which is not altered by deformation. The relaxation spectrum is therefore the basis for formulating constitutive equations in general. However, eqn (5.13) in this simple form is limited to very small strains. For example, the critical shear strain in a polyethylene melt is about 0·1 and in a suspension about 10^{-4}.

5.2.4. Extension of Linear Viscoelastic Equation to Large Strains

The general equation of linear viscoelasticity, eqn 5.13 is restricted to small strains for two reasons. Firstly, it describes the relaxation modes of the macromolecules in the equilibrium state and this equilibrium would be disturbed by large strains. Secondly, the strain measure which is commonly used neglects the quadratic terms compared to the linear ones. At large strains, this strain tensor fails in ways which will not be discussed here (see for instance Refs 2 or 3). It is interesting to note that for many materials the second strain criterion fails at much smaller strains than the first (material structure) criterion. If we simply replace the linearized strain tensor with a complete strain tensor, like the Finger strain tensor C^{-1}, then eqn (5.13) extends its validity to strains which may be (depending upon the specific polymer) increased up to two orders of magnitude. Such an equation has been derived by Lodge[11] from first principles. It is often called the 'Lodge Elastic Liquid' equation

$$\sigma(t) = \int_{-\infty}^{t} \sum_{i=1}^{N} \frac{g_i}{\lambda_i} e^{-(t-t')/\lambda_i} \mathbf{C}^{-1}(t', t) \, dt' \tag{5.20}$$

The integration has to follow the path of the material element. Components of $\mathbf{C}^{-1}$ are given in the Appendix C for shear and extension).

The Lodge Elastic Liquid can be written as a sum of N differential equations. In that formulation, the ith relaxation mode (g_i, λ_i) gives rise to a stress contribution, $\sigma_i(t)$, which depends on the local velocity gradient and the instantaneous rate of strain, $\mathbf{D}(t)$.

$$\sigma_i(t) + \lambda_i \hat{\sigma}_i(t) = g_i \lambda_i 2\mathbf{D}(t) \tag{5.21}$$

The total stress is assumed to be a linear superposition of the individual stresses of the N modes

$$\sigma(t) = \sum_{i=1}^{N} \sigma_i(t) \tag{5.22}$$

The time derivative in eqn (5.21) is the 'upper convected time derivative' of Oldroyd:[2]

$$\hat{\sigma}(t) = \dot{\sigma}(t) - \nabla \mathbf{v}^{\mathrm{T}} \cdot \sigma(t) - \sigma(t) \cdot \nabla \mathbf{v} \tag{5.23}$$

Computer simulations have primarily used the differential form of the constitutive equation[1] since the local stress is fully defined by the local velocity field and its gradient. Difficulties arise from the superposition of the N relaxation modes and from the limited stability (of Hadamard type) of the numerical codes. The integral constitutive equation, on the other hand, has the advantage that it easily accommodates the entire relaxation time spectrum. However, it is difficult to use in a numerical simulation since it requires the tracking of material elements along their path lines.[13-15]

5.2.4.1. Non-isothermal Flow
Processing flows are rarely isothermal and the temperature variations severely influence the stress. Hopkins[16] and Moreland and Lee[17] suggested a simple modification of the memory integral equations in order to account for non-isothermal histories, $T(t') \neq T_0$:

$$\sigma(t) = \int_{-\infty}^{t} \sum_{i=1}^{N} \frac{g_i}{\lambda_i} e^{-f(t',t)/\lambda_i} \mathbf{C}^{-1}(t', t) \, dt' \tag{5.24}$$

with an exponent

$$f(t', t) = \int_{t'}^{t} dt'' / a_T(t'') \qquad (5.25)$$

The temperature shift factor changes along the path line according to the temperature history, $a_T(t'') = a_T(T_0, T(t''))$. The non-isothermal effects result in a stretching or compressing of the time scale. Conveniently, one uses $f(t, t')$ as the new time. It should be noted that eqns (5.24) and (5.25) only apply to polymers for which the time–temperature superposition principle[9] holds.

For differential models, one evaluates the material parameters, g_i, λ_i at the instantaneous temperature. The procedure, however, is only valid for homogeneous temperature fields. For transient temperatures, $T(t, \mathbf{x})$, the differential models lose any advantage they otherwise might have for numerical simulations.

5.2.5. Finite Viscoelasticity

In the previous chapters, the rheological behavior was completely described by the linear relaxation time spectrum. Non-linear effects at large strain originated from the non-linearity of the strain tensor or from the Oldroyd time derivative (depending on the choice of integral or differential model). However, we must somehow account for the increased molecular mobility which, for instance, results in the shear-thinning phenomenon (see Section 5.2.2).

For a long time the increased molecular mobility of high rates of strain has led to the belief that the constitutive equations had to be modified with the rate of strain as the main parameter. However, all such equations failed when attempting to predict the dynamic mechanical behavior at high rate or when predicting the stress response to a sudden strain (step-strain experiment). The experiments of Einaga et al.[18] then showed that the characteristic parameter is not the rate of strain but the magnitude of the strain. This is concluded from measurements of the relaxing stress after a sudden strain (see Fig. 5.3). This time-dependent stress divided by the constant strain is equal to the relaxation modulus, $G(t)$. At small strain, the relaxation modulus is identical with the relaxation modulus which one may find from other linear viscoelastic experiments (such as the dynamic mechanical experiment as shown by Laun.[19]) At large strain, the relaxation modulus is simply shifted without changing shape

$$G(t, \gamma) = h(\gamma)\overset{\circ}{G}(t) \qquad (5.26)$$

The strain and the time dependence are given as the product of two independent functions. This result seems to have motivated several constitutive equations, especially the Doi–Edwards model and the factorized memory integral, eqn (5.27).

Doi and Edwards[20,21] and Curtiss and Bird[22] derived constitutive equations for polymeric liquids at large strain. These derivations start out with a model of the motion of linear macromolecules and result in an expression for the stress. Their equations will probably find much application in future modelling of polymer processing. However, they are still very difficult to use and will not be covered here. Kaye[23] and Bernstein et al.[24] proposed a model (the K-BKZ model) which is based on continuum mechanics arguments. It is nicely summarized by Larson,[3] and the reader is referred to these sources for further detail.

In the short space available, we will introduce two constitutive equations which were widely used for modelling the flow of polymeric liquids. These are the factorized memory integral model,[25] and the Giesekus model.

5.2.5.1. Factorized K-BKZ Model

A special case of the K-BKZ model has been proposed by Bogue and Doughty[25] in which the memory function was factorized into a time function and a strain function. However, they did not pursue this model further since comparison with the very limited experimental data of that time gave only marginal agreement between experiment and theory. Tanner and Williams,[26] Yen and McIntire,[27] Wagner,[28] and Osaki[29] continued to work with the model; the experiments of Einaga et al.[18] especially supported such an approach. The consequences of the model have been explored extensively by Wagner and it is therefore often called the 'Wagner model'. In this model, the time- and strain-dependence are factorized as suggested by Einaga et al.'s experiment:

$$\sigma(t) = \int\limits_{-\infty}^{t} \sum_{i=1}^{N} \frac{g_i}{\lambda_i} e^{-(t-t')/\lambda_i} h(t', t) \mathbf{C}^{-1}(t', t) \, dt' \qquad (5.27)$$

The integration has to follow the path of the material element.

The strain function, $h(t', t)$, has been measured in well-defined experiments. It seems to be a different function in shear, uniaxial extension, and in mixed flows. In this sense, eqn (5.27) is not a constitutive equation. The most simple form[30]

$$h = e^{-m\gamma} \quad \text{with} \quad \gamma(t', t) = \int_{t'}^{t} \dot{\gamma}(t'') \, dt'' \tag{5.28}$$

is already surprisingly accurate for predicting the shear stress and the first normal stress difference in transient and steady shear flows. Linear viscoelasticity is recovered at small strain, i.e. $h = 1$ for $\gamma \to 0$. A better fit is achieved with a sum of two such exponentials.[29,19] Uniaxial extension, together with shear, may be covered by an invariant form[30]

$$h = e^{\sqrt{\alpha \mathrm{I} + (1-\alpha)\mathrm{II} - 3}}, \quad 0 < \alpha < 1 \tag{5.29}$$

with the strain invariants

$$\mathrm{I} = \operatorname{tr} \mathbf{C}^{-1}(t', t) \quad \text{and} \quad \mathrm{II} = \{[\operatorname{tr} \mathbf{C}^{-1}(t', t)]^2 - \operatorname{tr}[\mathbf{C}^{-1}(t', t)]^2\}/2 \tag{5.30}$$

Equibiaxial extension[31] might also be described in this format, but it would require another α-value. The simplicity of eqn (5.27) is appealing for numerical simulations and should advance the modelling of polymer processing even if the equation is restricted to special types of flow. Temperature dependence can be included in the same way as suggested by Moreland and Lee.[17]

5.2.5.2. Giesekus Model

Giesekus[32,33] derived a constitutive equation from molecular arguments, in which the stress is squared

$$\boldsymbol{\sigma}(t) \cdot [1 + \beta \boldsymbol{\sigma}(t)/g] + \lambda \hat{\boldsymbol{\sigma}}(t) = g\lambda \, 2\mathbf{D}(t) \tag{5.31}$$

Parameter β ranges between 0 and 1. At small strain, and hence at small stress, the square term $\beta \boldsymbol{\sigma} \cdot \boldsymbol{\sigma}/g$ vanishes compared to the linear term, $\boldsymbol{\sigma}$, and linear viscoelastic behavior is recovered. Obviously, eqn (5.31) only shows the stress contribution of a single Giesekus mode with a modulus g and a time constant λ. The total stress is then predicted as linear superposition of all modes as discussed before, eqn (5.22).

The Giesekus model is one of the most simple viscoelastic constitutive equations. It describes linear and finite viscoelasticity surprisingly well and it has been successfully applied to the modelling of viscoelastic flows. In the special case of $\beta = 1/2$ (and for plane flows), it is equivalent to the Leonov model[34] which has also found application in processing models.

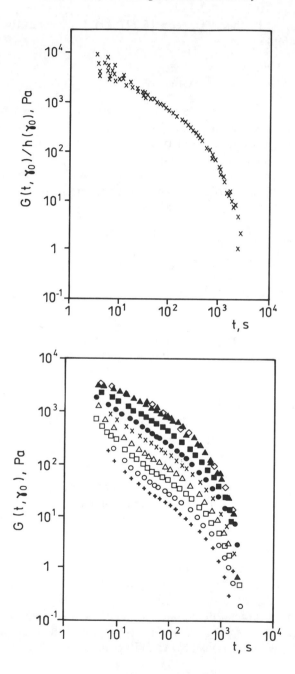

5.3. DETERMINATION OF MATERIAL PARAMETERS

5.3.1. Objectives of Computer-Aided Rheometry (CAR)

Our current CAR-program emphasizes three topics: time–temperature superposition, calculation of relaxation time spectra, and model calculations (including a crosscheck of material functions). The determination of the relaxation time spectrum assumes most attention since the spectrum is so important for rheological modelling, linear or non-linear. For a long time, the determination of the spectrum was considered a very difficult problem, and now it is performed by standard procedure.

In the following we will assume that carefully measured data are available. With the current rheometers this is giving us the steady shear viscosity $\eta(\dot{\gamma}, T_i)$, the components of the complex modulus G', $G''(\omega, T_i)$, and transient shear stress and normal stress over a wide range of times, rates ($\dot{\gamma}$ or ω), and at a discrete set of temperatures, T_i. The data will be first transformed into a *standard file format* so that it does not matter which instrument was used for obtaining the data or which literature source supplied the data file. The introduction of this data standard is essential for the following work.

With the above procedure, we propose to separate the data evaluation from the rheological experiment, i.e. we will not attempt a dialog between the evaluation software and the rheometer. However, the data evaluation will give much information about the reliability and range of applicability of the data. CAR has been proven to be an important tool for crosschecking between the various rheological experiments. Such knowledge will help to reduce the systematic error in the rheological experiments. This is very important since the systematic error is often very large, even if the stochastic error (noise) is negligibly small.

All of these tasks involve many repetitive steps which will be performed by the computer. The application of experimental data becomes very easy now, as demonstrated at the end of this chapter. Most calculations below apply to the linear viscoelastic range. An exception is the viscosity function. Software for large-strain predictions is in progress but we do not discuss it here.

Fig. 5.3. Apparent shear relaxation modulus $G(t, \gamma_0)$ for a 20% solution of polystyrene ($M_w = 1 \cdot 80\ 10^6$, $M_w/M_n = 1 \cdot 25$) in chlorinated diphenyl (Acrolor 1248) at $33 \cdot 5°C$. Values of strain γ_0 in the step strain experiment are $0 \cdot 41$ (▲), $1 \cdot 87$ (◇), $3 \cdot 34$ (■), $6 \cdot 68$ (×), $10 \cdot 0$ (△), $13 \cdot 4$ (□), $18 \cdot 7$ (○) and $25 \cdot 4$ (+). Vertical shifting gives a master curve for times bigger than $0 \cdot 1$ s. Ref.: Einaga et al. 1971.

5.3.2. Determination of Temperature Shift Factors

The experimental time (or frequency) window is restricted by the transducer characteristics of the rheometer. The experimental window is commonly enlarged by performing isothermal experiments at a range of discrete temperatures, and by applying time–temperature superposition. For this reason, this is the first important step of the data evaluation procedure. Temperature shift factors a_T and b_T can be determined experimentally from either linear viscoelastic or steady shear rheological experiments. In both cases isothermal experiments must be performed at discrete temperatures. Linear viscoelastic experiments can either be performed in the frequency or time domain. The dynamic moduli, $G'(\omega, T)$ and $G''(\omega, T)$, are measured in the frequency domain, and either the stress relaxation modulus, $G(t, T)$, or the creep compliance, $J(t, T)$ is measured in the time domain. For steady shear experiments the shear stress, $\sigma(\dot{\gamma}, T)$ is measured as a function of shear rate. The temperature shift factors can be determined from any of these quantities, but certain types of experiment may be preferred, depending upon the method that is used to determine the shift factors. In each case, the aforementioned quantities are plotted against either frequency (or rate) or time, depending on the experiment.

Once the experimental data have been plotted on log–log axes, the resultant curves can be shifted both horizontally and vertically to determine a_T and b_T respectively. For this purpose, one first chooses a reference curve at a reference temperature T_0 and then shifts each of the individual curves to overlay onto the reference curve. Individual paper plots at each temperature and a light box have been typically used to manually shift curves to obtain these shift factors. These tasks can be easily automated or performed interactively on the computer screen.

The automatic determination of shift factors using computer methods involves techniques that recognize and compare curve shapes. Shapes may be quantified by polynomial fitting of the data curves. The subsequent shifting of these fit curves onto the reference curve results in the desired master curve. For well-behaved data sets, very accurate curve fits have been obtained and automatic shifting is successful. However, many times, due to experimental scatter arising from instrument measuring limitations and other factors, the data curves may not be easily fit and shifted based on mathematical algorithms. In these cases, the curves can be shifted more easily, and to a very good degree, by the human eye. The key here is to provide an interactive curve-shifting capability on screen. There the user performs the shifting, sees the fitted results, and modifies the shifted curves until the curves are overlaid on the reference curve.

Certain types of experiment may provide easier determination of shift factors. For instance, in steady shear-flow experiments, the shear stress at high rates tends to be fairly flat, and shifting horizontally to determine a_T can be prone to significant errors. For polymer melts, linear viscoelastic dynamic modulus measurements provide the best data for determining the shift factors. The experiments require small samples, are performed very quickly and the resultant $G'(\omega, T)$ and $G''(\omega, T)$ curves provide two curves of varying shapes that can be shifted simultaneously. Another advantage is that one can easily determine a_T first by shifting the tan δ (ratio: G''/G') curves, in which the vertical shift dependency is removed because both G' and G'' have the same b_T.

A typical shifting procedure is shown in the following figures for poly-(dimethyl siloxane) (PDMS). Dynamic moduli and tan δ curves, obtained at five different temperatures, are shown in Figs 5.4–5.6. The data will be reduced to a reference temperature of 60°C. The values of a_T were determined first by plotting G', G'' and tan δ simultaneously and shifting each set of curves horizontally until all the tan δ curves overlay on the 60°C curve, as shown in Fig. 5.7. As seen in Fig. 5.7, the G' and G'' curves do not overlay at this point because they have not been shifted vertically yet (all $b_T = 1$). To determine b_T, each set of curves is then shifted vertically, until all the G' and G'' curves overlay the 60°C curve as shown in Fig. 5.8. The shift factors determined for this polymer are listed in Table 5.1.

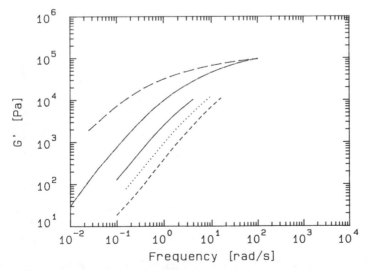

Fig. 5.4. Storage modulus, G', of PDMS *versus* frequency; (——) −60°C, (— · —) 0°C, (——) 60°C, (· · · · ·) 120°C, (– – – –) 180°C.

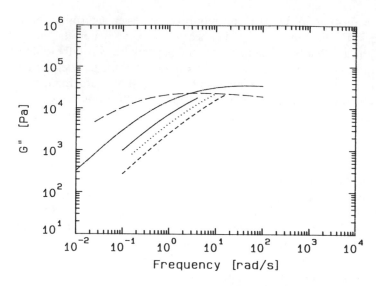

Fig. 5.5. Loss modulus, G'', of PDMS *versus* frequency; (——) −60°C, (— · —) 0°C, (——) 60°C, (· · · · ·) 120°C, (− − − −) 180°C.

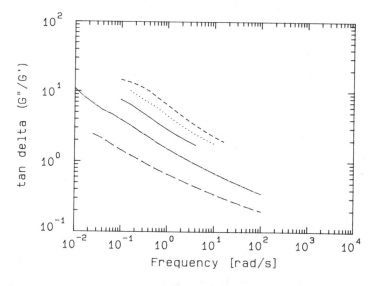

Fig. 5.6. Tan δ of PDMS *versus* frequency; (——) −60°C, (— · —) 0°C, (——) 60°C, (· · · · ·) 120°C, (− − − −) 180°C.

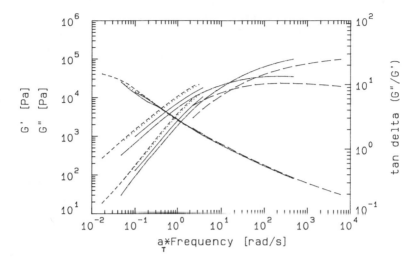

Fig. 5.7. Dynamic data of PDMS with tan δ shifted horizontally to 60°C. All $b_T = 1$. (———) −60°C, (— · —) 0°C, (———) 60°C, (· · · · ·) 120°C, (– – – –) 180°C.

Once a_T and b_T have been obtained graphically the shift factors are fit to the WLF and Arrhenius equations for a_T and an Arrhenius equation for b_T, as a function of temperature. As mentioned in the previous sections, the WLF equation typically describes a_T very well at temperatures near T_g up to about $T_g + 100$ K, and the Arrhenius equation is essentially a two-parameter fit (E_a and T_0) whereas the WLF equation has three fitting parameters (C_1, C_2 and T_0). Because of the extra fitting parameter, the WLF equation can fit the data better in areas of higher curvature, such as near T_g. Non-linear regression techniques may be used to determine the parameters. Using these methods, all the parameter values are allowed to float until the residuals are minimized. Therefore, the reference temperature calculated from the non-linear regression is most likely different to the one initially chosen for curve shifting. If one holds the reference temperature constant (assuming that the rheological and temperature data are both accurate and precise at the reference temperature), then the Arrhenius and WLF equations can both be fit with linear regression techniques. The Arrhenius equation then becomes a single parameter fit and the WLF a two-parameter fit.

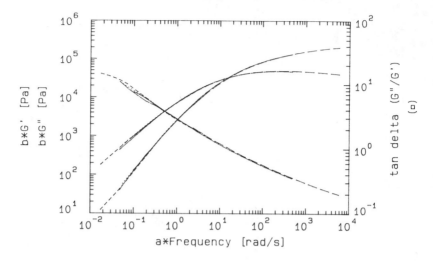

Fig. 5.8. Dynamic data of PDMS with tan δ shifted horizontally and G' and G'' shifted vertically to 60°C; (— —) −60°C, (— · —) 0°C, (———) 60°C, (· · · · ·) 120°C, (− − − −) 180°C.

TABLE 5.1
Horizontal and Vertical Temperature Shift
Factors of PDMS

Temperature (°C)	log a_T	log b_T
−60	1·77	0·32
0	0·68	0·13
60	0·00	0·00
120	−0·45	−0·13
180	−0·76	−0·18

Reference temperature $T_0 = 60$°C.

The resultant a_T and b_T values, determined from the data shown in Fig. 5.8, appear in Fig. 5.9. Both a_T and b_T have a stronger dependence upon temperature as one gets closer to T_g ($T_g = -120$°C for PDMS). Also, a_T is seen to have a much greater dependence upon temperature than b_T. PDMS is one of the least temperature-dependent polymers, so most other commercial polymers will have larger E_a and C_1 values.

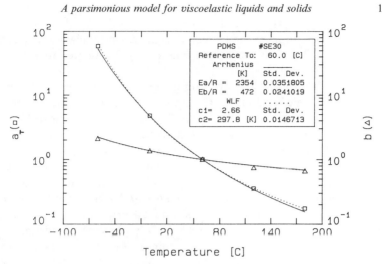

Fig. 5.9. Horizontal and vertical shift factors of PDMS. (——) Arrhenius fit, (· · · · ·) WLF fit.

Once these shift factors and reference temperatures are obtained, predictions of any linear viscoelastic function at any intermediate temperature can be calculated by the equations discussed in the earlier portions of this chapter. Time–temperature superposition is found with most polymeric solids and liquids. Exceptions are polymers which undergo some sort of structural transition (phase separation in blends and block copolymers, nematic/isotropic transition in liquid crystalline polymers, crystallization, ionic bonding, etc.).

5.3.3. Determination of Relaxation Time Spectrum

The relaxation modulus may be measured directly in the so-called 'step-strain' experiment. However, more advanced are dynamic mechanical methods in which the polymer is strained sinusoidally at a frequency ω (see Fig. 5.9). The measured stress response decomposes into an in-phase and an out-of-phase component: the storage modulus G' and the loss modulus G''. Using the discrete spectrum of eqn (5.14) together with eqn (5.13) the dynamic moduli, G' and G'', are calculated as

$$G'(\omega) = \sum_{i=1}^{N} g_i \frac{(\omega \lambda_i)^2}{1 + (\omega \lambda_1)^2} \tag{5.32}$$

$$G'(\omega) = \sum_{i=1}^{N} g_i \frac{\omega \lambda_i}{1 + (\omega \lambda_i)^2} \tag{5.33}$$

From these equations, it can be seen that the discrete relaxation spectrum, g_i, λ_i, $i = 1, 2, 3, \ldots, N$ may be determined by fitting measured values of the dynamic moduli.

Many different methods have been proposed for converting the readily available dynamic moduli G', G'' into the relaxation time spectrum, g_i and λ_i.[9,19,35] Baumgaertel and Winter[36] have recently developed a robust numerical method which converts dynamic mechanical data from the frequency to the time domain. The method avoids illposedness, which seems to be inherent in the curve fitting when choosing too many parameters. The method is called 'Parsimonious Modelling' since it attempts to find the spectrum with the *smallest number of Maxwell modes* which still represents the data within the experimental error margin. A closer fit could be achieved but it would not have any physical significance.

Required are input data G', G'' within a finite frequency window. A non-linear regression simultaneously adjusts g_i, λ_i (*both* have to be treated as freely adjustable material parameters!) to obtain a best fit of G', G''.

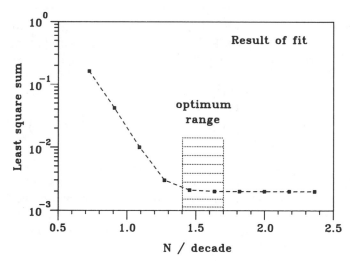

Fig. 5.10. Parsimonious model: Dependence of the quality of the fit on the number of relaxation modes per decade of dynamic data. The least square deviation between data and fit is poor if too few relaxation modes are chosen. The fit improves dramatically as N increases and, with surprisingly few modes per decade, represents the data within experimental error. Above a certain high number of modes, the fit does not improve much further and the problem becomes ill posed. Obviously, there is an optimum range of relaxation modes which depends upon the material under investigation and is not known *a priori*.

It automatically chooses a suitable set of initial values $g_{i,0}$, $\lambda_{i,0}$. The number of relaxation times adjusts during the iterative calculations, depending on the needs for improved fit (Fig. 5.10). The method is especially useful when working with complex polymeric materials, since the data does not need to be extrapolated beyond the experimental frequency window (as with a Fourier Transform), and no empirical correlations are necessary. The calculated relaxation spectrum may be directly converted into the retardation spectrum, j_i, Λ_i, $i = 1, 2, \ldots, (N-1)$ as reported by Baumgaertel and Winter.[36] The calculated discrete set of parameters g_i and λ_i (also j_i and Λ_i) is, obviously, only valid in a time window which corresponds to the frequency window of the input data, $t_{min} = 1/\omega_{max}$ and $t_{max} = 1/\omega_{min}$. However, in cases where the terminal time of the material is known (longest relaxation mode for liquids or equilibrium modulus for solids), the upper time-limit becomes irrelevant and in turn the discrete spectra become valid for infinitely long times.

5.3.3.1. Typical Spectra of Polymers

Figures 5.11 and 5.12 show typical dynamic mechanical data and the corresponding discrete relaxation and retardation spectra of PDMS at

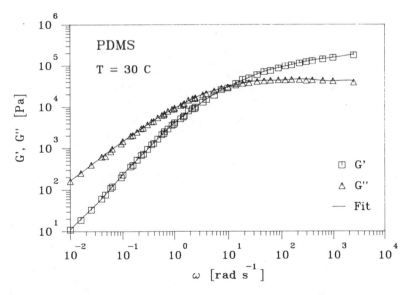

Fig. 5.11. Dynamic moduli *versus* frequency of PDMS (master curve after time temperature superposition).

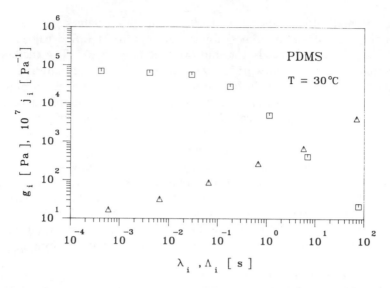

Fig. 5.12. Discrete relaxation (□) and retardation (Δ) spectra calculated numerically from data of Fig. 5.11.

the reference temperature $T = 30°C$. The line through the data points is calculated with the fitted relaxation spectrum of Table 5.2. The calculated relaxation modulus and the creep compliance are plotted in Fig. 5.13.

A second example shows the evolution of the relaxation modulus of a crosslinking polymer that undergoes gelation. The dynamic mechanical

TABLE 5.2
Discrete Relaxation and Retardation Spectra of PDMS

i	g_i (Pa)	λ_i (s)	j_i (Pa^{-1})	Λ_i (s)
1	$7 \cdot 093 \times 10^4$	$4 \cdot 180 \times 10^{-4}$	$1 \cdot 755 \times 10^{-6}$	$5 \cdot 964 \times 10^{-4}$
2	$6 \cdot 325 \times 10^4$	$4 \cdot 056 \times 10^{-3}$	$3 \cdot 316 \times 10^{-6}$	$6 \cdot 530 \times 10^{-3}$
3	$5 \cdot 710 \times 10^4$	$2 \cdot 885 \times 10^{-2}$	$8 \cdot 944 \times 10^{-6}$	$6 \cdot 518 \times 10^{-2}$
4	$2 \cdot 817 \times 10^4$	$1 \cdot 796 \times 10^{-1}$	$2 \cdot 778 \times 10^{-5}$	$6 \cdot 782 \times 10^{-1}$
5	$5 \cdot 024 \times 10^3$	$1 \cdot 127 \times 10^0$	$6 \cdot 988 \times 10^{-5}$	$5 \cdot 774 \times 10^0$
6	$4 \cdot 250 \times 10^2$	$6 \cdot 994 \times 10^0$	$4 \cdot 145 \times 10^{-4}$	$6 \cdot 994 \times 10^1$
7	$2 \cdot 193 \times 10^1$	$7 \cdot 732 \times 10^1$		

Reference temperature $T_0 = 30°C$.

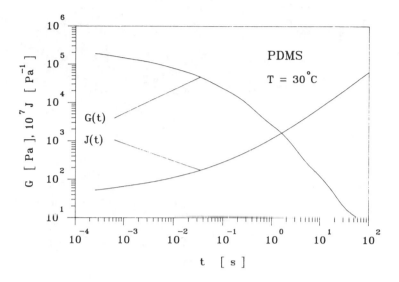

Fig. 5.13. Relaxation modulus, $G(t)$, and creep compliance, $J(t)$, calculated with discrete spectra of Fig. 5.12.

data (Fig. 5.14) of Venkataraman[37] are converted into the relaxation modulus (Fig. 5.15). The evolution of the relaxation modulus can be divided into two regions: before the gel point, when the polymer can relax completely, and beyond the gel point, when the polymer can only relax to an equilibrium value (equilibrium modulus, G_e). At the gel point the polymer relaxes in a power law (see eqn (5.3)).

5.3.3.2. Discrete Spectra Extension using Transient Experimental Data

In cases where time–temperature superposition is not possible, the time window is rather small (0·01–100 s). However, if transient experimental data are available, i.e. from a step-strain or a creep experiment, the discrete spectra can often be extended to longer times (Fig. 5.16).

In cases where the relaxation modulus is known (from a step-strain experiment ($t'_{max} > t_{max}$), relaxation modes with longer relaxation times (if existent) are determined by a non-linear regression from eqn (5.14) within the new time window (t_{min}–t'_{max}).

The same procedure applies to data from a creep experiment. Additional modes in the range of longer times can be determined by a

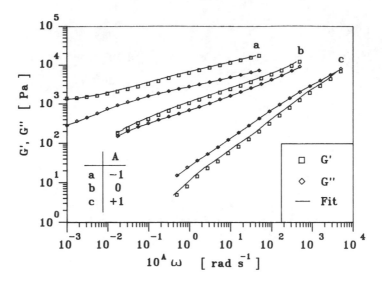

Fig. 5.14. Evolution of the dynamic moduli of a crosslinking polymer at increasing degree of crosslinking (increased crosslinking time). The lines are calculated with discrete spectra. Data in the liquid and in the solid range are approximated equally well. No attempt is made to extrapolate to the equilibrium modulus.

non-linear regression from

$$J(t) = J_e + t/\eta_0 + \sum_{i=1}^{N-1} j_i(1 - e^{-t/\Lambda_i}) \tag{5.34}$$

within the new time window ($t_{min}-t_{max}$). J_e and η_0 are material parameters.

5.3.4. Applications for Linear Spectra

5.3.4.1. Simulation of transient experiments using discrete spectra
The discrete spectra are very useful in exploring constitutive equations and in some cases even analytical solutions are obtained (see Laun,[19] for instance). Non-linear responses can easily be calculated if an appropriate model is available. However, linear viscoelastic responses can be expressed in analytic form by integrating eqn (5.13) for specific strain histories such as start-up of shear flow or creep.

For example, in a *stress growth and relaxation* experiment, a material is sheared initially with a constant strain rate, $\dot{\gamma}_0$, for a time t_0. The strain is then kept at a constant value. During this experiment, the stress is

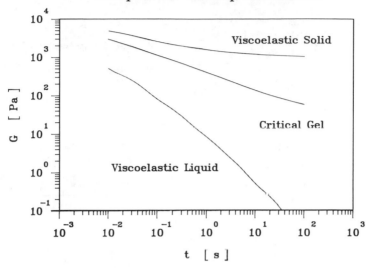

Fig. 5.15. Evolution of relaxation modulus during liquid/solid transition as calculated from Fig. 5.14.

recorded. In the linear viscoelastic region, the stress response during the initial shear, $0 \leq t \leq t_0$, is given by

$$\sigma(t) = \dot{\gamma}_0 \left(G_0 t + \sum_{i=1}^{N} g_i \lambda_i (1 - e^{-t/\lambda_i}) \right)$$

$$N_1(t) = \dot{\gamma}_0^2 \left(G_0 t^2 + 2 \sum_{i=1}^{N} g_i \lambda_i^2 [1 - (1 + t/\lambda_i) e^{-t/\lambda_i}] \right) \qquad (5.35)$$

and during stress relaxation $(t_0 < t)$

$$\sigma(t) = \dot{\gamma}_0 \left(G_0 t_0 + \sum_{i=1}^{N} g_i \lambda_i (e^{-(t-t_0)/\lambda_i} - e^{-t/\lambda_i}) \right)$$

$$N_1(t) = \dot{\gamma}_0^2 \left(G_0 t_0^2 + 2 \sum_{i=1}^{N} g_i \lambda_i^2 [1 - (1 + t_0/\lambda_i) e^{-t_0/\lambda_i}] e^{-(t-t_0)/\lambda_i} \right) \qquad (5.36)$$

These predictions may be compared with experimental shear stress data as shown in Fig. 5.17 for a liquid, Fig. 5.18 for a gel and Fig. 5.19 for a solid. The agreement is acceptable in this case.

Parameter identification

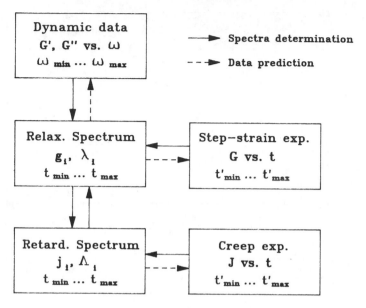

Fig. 5.16. Data flow sheet: First, the discrete spectra are determined from dynamic moduli. The time window of validity can be extended if either their relaxation modulus, $G(t)$, or the creep compliance, $J(t)$, are known from transient experiments.

A second very common experiment involves *creep* and *recovery*. Here the material is initially sheared under a constant stress, τ_0, for a time t_0 and the strain response is measured. In the linear viscoelastic region, the strain response during initial shear, $0 \leq t \leq t_0$, is given by

$$\gamma(t) = \tau_0 \left(J_e + t/\eta_0 + \sum_{i=1}^{N-1} j_i (1 - e^{-t/\Lambda_i}) \right) \tag{5.37}$$

and during recovery $(t_0 < t)$

$$\gamma(t) = \tau_0 \left(\tau_0/\eta_0 + \sum_{i=1}^{N-1} j_i (e^{-(t-t_0)/\Lambda_i} - e^{-t/\Lambda_i}) \right) \tag{5.38}$$

These predictions agree well with the experimental data shown in Fig. 5.20 for a liquid, Fig. 5.21 for a gel and Fig. 5.22 for a solid.

Viscoelastic Liquid

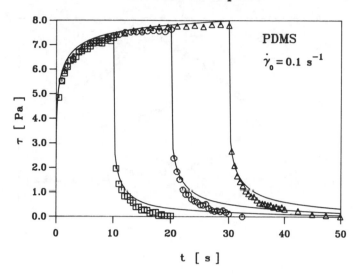

Fig. 5.17. Comparison of shear stress data from the stress growth and recovery experiment with PDMS at 30°C with corresponding simulation. The material before its gel point behaves as a viscoelastic liquid.

If shift factors are available, all simulations can easily be extended to any temperature within the accessible temperature window.

5.3.4.2. Estimation of Viscosity Value from Linear Relaxation Spectrum

In many instances, the steady shear viscosity is needed while only the relaxation time spectrum is available. Fortunately, for many polymers there exist heuristic relations which predict the viscosity from the linear relaxation time spectrum. The first one is the *Cox–Merz Relation*[38]

$$\eta(\gamma) = \eta^*(\omega)\bigg|_{\omega=\dot\gamma} = \frac{\sqrt{G'^2 + G''^2}}{\omega}\bigg|_{\omega=\dot\gamma}$$

$$= \left[\left[\sum_{i=1}^{N} \frac{g_i \dot\gamma_i \lambda_i^2}{1+(\lambda_i\dot\gamma)^2}\right]^2 + \left[\sum_{i=1}^{N} \frac{g_i \lambda_i}{1+(\lambda_i\dot\gamma)^2}\right]^2\right]^{1/2} \quad (5.39)$$

η^* is called the dynamic viscosity. The Cox–Merz relation can convert dynamic mechanical data, G', G'', directly into the viscosity function or,

Critical Gel

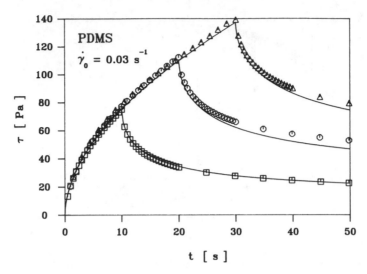

Fig. 5.18. Comparison of shear stress data from the stress growth and recovery experiment at 30°C with corresponding simulation. At the gel point, power law relaxation is observed.

if the spectrum is available, the parameters g_i/λ_i may be used in the above formula. The parameter values are taken at temperature T.

The second relation has been suggested by Gleissle.[39] It may be expressed as

$$\eta(\dot{\gamma}, T) = \sum_{i=1}^{N} g_i\lambda_i(1 - e^{-1/(\dot{\gamma}\lambda_i)}) \qquad (5.40)$$

The Gleissle mirror relation is easier to use for calculating the steady shear viscosity from a known relaxation time spectrum, while the Cox–Merz relation is more useful for direct plotting of the viscosity function from a plot of the dynamic viscosity, η^*.

5.4. CONCLUSIONS

An experiment with a rheometer is only the first step in rheological characterization. Data processing is as important when it comes to the critical evaluation of noisiness,[40] the range of validity, and the determination

Viscoelastic Solid

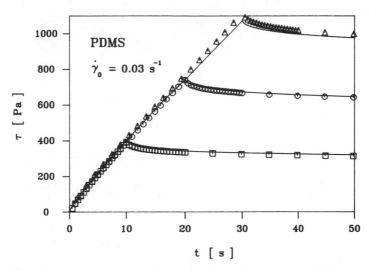

Fig. 5.19. Comparison of shear stress data from the stress growth and recovery experiment at 30°C with corresponding simulation. The material after its gel point behaves as a viscoelastic solid.

of the relaxation modes which have to be known when solving applied problems. A computer package has been developed for this purpose. It accepts data files from a large variety of rheometers, superimposes and shifts them, and allows comparison of a variety of different transient modes in a computer-aided modelling process. Only after having scrutinized the data systematically and having found consistency within a data set, do we consider the rheological characterization acceptable. With the current analysis, agreement between various measurements is only expected within the linear viscoelastic region, while deviations from linear viscoelasticity are clearly identifiable in the computer graphs. The analysis of non-linear viscoelastic data could follow a similar procedure as described here. However, it has not yet been implemented in the CAR code.

An important outcome of the computer-aided rheometry is the condensation of the rheological behavior in a small set of material parameters, the parsimonious model. Such a model can be readily used in flow calculations, and it can be stored in a data base where it is easily accessed during future applications.

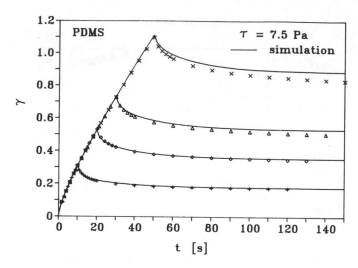

Fig. 5.20. Comparison of experimental data from creep recoil experiment at 30°C with corresponding simulation. The material before its gel point behaves as a viscoelastic liquid.

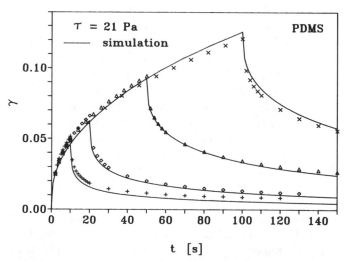

Fig. 5.21. Comparison of experimental data from creep recoil experiment with corresponding simulation. At the gel point, power law relaxation is observed.

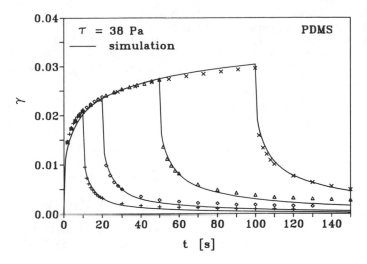

Fig. 5.22. Comparison of experimental data from creep recoil experiment with corresponding simulation. The material after its gel point behaves as a viscoelastic solid.

REFERENCES

1. M. J. Crochet, A. R. Davies and K. Walters, *Numerical Simulation of Non-Newtonian Flow*, Elsevier, Amsterdam, 1984.
2. R. B. Bird, R. C. Armstrong and O. Hassager, *Dynamics Of Polymeric Liquids*, John Wiley, New York, 1987.
3. R. G. Larson, *Constitutive Equations For Polymer Melts And Solutions*, Butterworths, Boston, 1988.
4. H. H. Winter and F. Chambon, *J. Rheol.* 1986, **30**, 367.
5. F. Chambon and H. H. Winter, 1990, *J. Rheol.*, 1987, **31**, 683.
6. S. Guskey and H. H. Winter, 1990, *J. Rheol.*, 1991, **35**, 1191.
7. S. Middleman, *Fundamentals Of Polymer Processing*, McGraw-Hill, New York, 1977.
8. C. J. Rauwendaal, *Polymer Extrusion*, Hanser, München, 1986.
9. J. D. Ferry, *Viscoelastic Properties Of Polymers*, John Wiley, New York, 1980.
10. P. J. Carreau, *Trans. Soc. Rheol.*, 1972, **16**, 99.
11. A. S. Lodge, *Elastic Liquids*, Academic Press, New York, 1964.
12. J. G. Oldroyd, *Proc. Roy. Soc.*, 1950, 1977, 426.
13. H. H. Winter, *J. Non-Newtonian Fluid Mech.*, 1982, **10**, 157; K. H. Wei, M. E. Nordberg and H. H. Winter, *Polym. Eng. Sci.* 1987, **27**, 1390.
14. S. Dupont, J. M. Marchal and M. J. Crochet, *J. Non-Newtonian Fluid Mech.*, 1985, **17**, 157.

15. K. Adachi, *Rheol. Acta*, 1986, **25**, 555.
16. I. L. Hopkins, *J. Polym. Sci.*, 1958, **28**, 631.
17. L. W. Moreland and E. H. Lee, *Trans. Soc. Rheol.*, 1960, **IV**, 233.
18. Y. Einaga, K. Osaki, M. Kurata, S. Kimura and M. Tamura, *Polym. J.*, 1971, **2**, 550; M. Fukuda, K. Osaki and M. Kurata, *J. Polym. Sci., Polym. Phys. Ed.*, 1975, **13**, 1563.
19. H. M. Laun, *Rheol. Acta*, 1978, **17**, 1.
20. M. Doi and S. F. Edwards, *J. Chem. Soc. Faraday Trans. II*, 1978, **74**, 1789; 1978, **74**, 1802; 1978, **74**, 1818.
21. M. Doi and S. F. Edwards, *J. Chem. Soc. Faraday. Trans. II*, 1979, **75**, 88.
22. C. F. Curtiss and R. B. Bird, *J. Chem. Phys.*, 1981, **74**, 2016; 1981, **74**, 2026.
23. A. Kaye, UK Note 134, College of Aeronautics, Cranfield, 1962.
24. B. Bernstein, E. A. Kearsley and L. J. Zapas, *Trans. Soc. Rheol.*, 1963, **7**, 391.
25. D. C. Bogue and J. O. Doughty, *I. and E. Fundamentals*, 1966, **5**, 243; 1967, **6**, 388.
26. R. I. Tanner and G. Williams, *Trans. Soc. Rheol.*, 1970, **14**(1), 19.
27. H. C. Yen and L. V. McIntyre, *Trans. Soc. Rheol.*, 1974, **18**(2), 495.
28. M. Wagner, *Rheol. Acta.*, 1976, **15**, 136.
29. K. Osaki, *Proc. 7th Int. Congr. on Rheology*, Gothenberg, Sweden, 1976, p. 104.
30. M. Wagner, *J. Non-Newtonian Fluid Mech.*, 1978, **4**, 39.
31. P. Soskey and H. H. Winter, *J. Rheol.*, 1985, **29**, 493.
32. H. W. Giesekus, *Rheol. Acta*, 1966, **5**.
33. H. W. Giesekus, *J. Non-Newtonian Fluid Mech.*, 1982, **11**, 69.
34. A. I. Leonov, *Rheol. Acta*, 1976, **15**, 85.
35. G. Friedrich and B. Hoffmann, *Rheol. Acta*, 1983, **22**, 425.
36. M. Baumgaertel and H. H. Winter, *Rheol. Acta*, 1989, **28**, 511.
37. S. Venkataraman, PhD Thesis, University of Massachusetts, 1988.
38. W. P. Cox and E. H. Merz, *J. Polym. Sci.*, 1958, **28**, 619.
39. W. Gleissle, Doctoral Thesis, University of Karlsruhe, 1978; for Mathematical Formulation see H. H. Winter, *Polym. Eng. Sci.*, 1980, **20**, 406.
40. M. Baumgaertel and H. H. Winter, *J. Non-Newtonian Fluid Mech.* 1992 (in print).

APPENDIX: DEFINITIONS OF STRESS AND STRAIN

Rate of Strain

The instantaneous rate of strain is defined with the velocity gradient

$$\mathbf{D}(t) = (\nabla \mathbf{v}^T + \nabla \mathbf{v})/2 \qquad (\text{A.1})$$

It has the Cartesian components

in shear:

$$\mathbf{D}(t) = \begin{vmatrix} 0 & \dot{\gamma}/2 & 0 \\ \dot{\gamma}/2 & 0 & 0 \\ 0 & 0 & 0 \end{vmatrix} \tag{A.2}$$

with $\dot{\gamma}(t) =$ shear rate

in extension

$$\mathbf{D}(t) = \begin{vmatrix} \dot{\varepsilon}_1 & 0 & 0 \\ 0 & \dot{\varepsilon}_2 & 0 \\ 0 & 0 & \dot{\varepsilon}_3 \end{vmatrix} \tag{A.3}$$

with $\dot{\varepsilon}_i(t) =$ extension rates

Finger Strain Tensor
The strain between two states, t' and t, is described by the Finger strain tensor, $\mathbf{C}^{-1}$. Its components in shear and extension are given by the integration of the rate of strain over the time interval from t' to t:

in shear:

$$\mathbf{C}^{-1}(t', t) = \begin{vmatrix} 1 + \gamma^2 & \gamma & 0 \\ \gamma & 1 & 0 \\ 0 & 0 & 1 \end{vmatrix} \tag{A.4}$$

with the shear strain

$$\gamma(t', t) = \int_{t'}^{t} \dot{\gamma}(t'') \, dt'' \tag{A.5}$$

in extension:

$$\mathbf{C}^{-1}(t', t) = \begin{vmatrix} L_1^2 & 0 & 0 \\ 0 & L_2^2 & 0 \\ 0 & 0 & L_3^2 \end{vmatrix}^2 \tag{A.6}$$

with the extensions

$$L_i(t', t) = \exp\left\{ \int_{t'}^{t} \dot{\varepsilon}_i(t'') \, dt'' \right\}$$ (A.7)

For complex flows, the components of $\mathbf{C}^{-1}$, are too complicated to be discussed here. Further details are discussed in the books by Bird *et al.*[2] and by Larson.[3]

Stress
Consider an infinitesimal material plane, dA, of unit normal $\mathbf{n}$ at position $\mathbf{x}$ in the material. The force per unit area, the traction $\mathbf{f}$, which is acting on dA is given by the stress tensor

$$\mathbf{f}(\mathbf{n}, \mathbf{x}) = \mathbf{n} \cdot \tau(\mathbf{x})$$ (A.8)

Effectively, the stress tensor is an operator which transforms the unit normal, $\mathbf{n}$, into the traction, $\mathbf{f}$.

The components of the stress tensor in a Cartesian coordinate system, δ,

$$\tau_{ij} = \delta_i \cdot \tau \cdot \delta_j$$ (A.9)

can be understood as components of a traction vector $\mathbf{f}^{(i)}$ on a plane of unit normal

$$\mathbf{n}^{(i)} = \delta_i$$ (A.10)

By definition, pulling is a positive normal traction and pressure is a negative normal traction on the material plane. The stress tensor τ is commonly decomposed into an isotropic component, the pressure p, and the extra stress σ

$$\tau = -p\mathbf{1} + \sigma$$ (A.11)

The extra stress is described in rheological constitutive equations.

ACKNOWLEDGEMENTS

We gratefully acknowledge the many helpful discussions with S. Venkataraman, C. G. Franco, M. Reuther, and K. Deutscher.

Chapter 6

Rheological Studies Using a Vibrating Probe Method

RICHARD A. PETHRICK

Department of Pure and Applied Chemistry, University of Strathclyde, Glasgow, UK

6.1. INTRODUCTION

The rheological properties of liquid media may be divided into two broad groupings: those that exhibit isochronal behaviour and those that do not.

Many liquids exhibit non-isochronal behaviour but are chemically stable, and if allowed sufficient time to recover will return to their equilibrium state. There are, however, media which do not fall into the above classification; these are systems in which the chemical structure of the fluid changes with time as a consequence of chemical reaction or of non-reversible physical processes. These systems are difficult to investigate, since they require that the measurement apparatus be capable of accurate and precise determination of rheological properties over large changes in magnitude. Within this group of systems are thermosetting resins, plastisols, thermoplastic modified thermosets and thermosetting composites and filled material. Traditionally, rheological measurements are performed in conditions of either constant stress or strain and geometrically involve cone and plate, capillary, coaxial cylinder or parallel-plate configurations.[1] These instruments are capable of providing well-defined shear fields and allow precise determination of the viscosity coefficients over a wide range of shear stresses and rates. One of the principle problems encountered with the use of these techniques for the characterization of reactive systems is that the cure process can lead ultimately to the generation of either a sticky gum or a vitrified sample. In either case the medium, once it is cured, sticks the elements of the rheometer together, leading to a costly cleaning or recovery exercise. Replacement of a cone and plate can cost around £200; which is usually an unacceptable price to pay for obtaining data on a reactive system. Recently, manufacturers have started to offer cheap throw-away elements which has reduced the cost of such studies. There is, however, the danger that many of the sensitive elements in the rheometer may be severely damaged during the measurement of the large viscosities encountered in the latter stages of cure. An alternative approach, which can be applied to such studies, is based on the *vibrating probe* method, and this is the main topic of this article.

The term *cure* is widely used in industry; however, although it means different things to different people, it is usually associated with the processes of polymerization, chain extension, irreversible physical entanglement or crosslinking. Such processes are characterized by an increase in the viscosity, leading to the formation of a stable three-dimensional gelled network characteristic of *cure* in many paints, powder coatings, composites and thermosetting filled systems. The point at which an infinite continuous matrix is formed is usually referred to as *gelation*, and is a characteristic property of such systems.[2-7] In approaching this state, the liquid will be transformed, through an intermediate state of polymer dissolved in monomer, to a gelling matrix plasticized by monomer to a vitrified glassy state. This process, or in the case of a drying polymer film, solvent loss, leads to an increase in viscosity with a value of approximately 10^4 Pa s being achieved at the gel point. The phenomenon of gelation is not uniquely defined, unlike a melt transition or a glass transition, but is theoretically the point at which the viscosity tends to infinity.[5-7]

Many of the traditional methods of assessing cure are based on a perception of when a material is dry to touch, demoulding becomes possible or the rigidity has passed a particular threshold as in the cocktail stick method. What emerges is a picture of cure which is very subjective, related to performance expectations, which are process-specific. In practice, the processor seeks information in terms of a number of quantities which are related to their application, but usually conform to the following common definitions.[8]

6.1.1. Application Time/Life, Pot Life, Pour Time

These terms are essentially the same and refer to the time available for the resin to be transferred from mixing vessel to the mould or site of application. The mode of transfer will depend upon the processes involved; pouring as in filling moulds, brushing when resin is applied to a substrate or fibre mat, and gun extrusion or trowelling when the resins are used as sealants or fillers. The appropriate level of viscosity depends upon which method is being applied, and leads to different perceived values for the same system in different applications. The definitions are usually subjective, but correspond to viscosities in the range 10^{-2} to 10^2 Pa s, and imply that a knowledge of the isothermal viscosity–time curve, in principle, should allow these quantities to be unambiguously defined.

6.1.2. Working Life/Time

This term usually refers to the period beyond which a mixture is incapable of being remoulded or reworked, and may be associated with the point at which a three-dimensional gel is formed; but it may be lower than the accepted value of 10^4 Pa s, if a distinct gel/sol phase is developed to any extent in the curing system.

6.1.3. Gel Time

This is synonymous with the point beyond which the mixture is incapable of flow and in molecular terms refers to the point at which an infinite network is formed.[6-7] With care, this point is capable of being reproducibly determined according to BS 2782, Part 8 Methods 835 A–D: 1980. Rheologically this process may be ascribed to the change from a liquid to a rubbery state.

6.1.4. Tack-Free Time, Demould Time

Technically, the development of a sufficiently robust state which is resistant to damage by contact or handling is critical in many moulding applications. For a sealant formulation this is the tack-free time, i.e. the time interval before the sealant can resist damage by touch or from settling dirt or rain. For moulded resins or cast elastomers it is the demould time when the semi-cured article may be safely removed to free the mould for the next productive cycle. Methods for the determination of tack-free time are contained within ASTM C679-87 and D2377-84.

6.1.5. Cure Time

This is generally accepted as the time required for the completion of the reaction process and can be calculated theoretically from the rate constants for the reaction. In practice, the cure time may be determined from experimental measurement of the degree of conversion using differential scanning calorimetry (DSC) or, alternatively, infrared spectroscopy. Both observations are concerned with determination of the extent of reaction either using the loss in intensity of a band characteristic of the monomer state or the exotherm related to the residual monomer. A more pragmatic definition relates cure development to a property of the system, such as the tensile strength or hardness. The desired end point or cure time is when the optimum property is achieved. This may be alternatively assessed using observation of a shift in the glass transition using techniques such as thermally stimulated discharge TSD of an electret.

6.2. QUALITY CONTROL METHODS

The ability to determine each of the above quantities is critical in terms of quality control,[8] trouble shooting, smart processing[9,10] and production development. An instrument applicable to monitoring the curing characteristics of systems as diverse as curable liquid types must be capable of covering an effective range from 10^{-2} to 10^{6} Pa s, if not higher. Such a requirement is generally speaking beyond the capability of the conventional viscometers,[1] but is, however, met by vibrating probe instruments.

6.2.1. Definition of the Curing Process

A number of attempts have been made to simplify the definition of the cure process and one of the most widely accepted is that based on the time–temperature-transformation (TTT) diagram,[11–14] Fig. 6.1, which relates the various states of matter encountered (i.e. liquid; sol/gel and gel/rubber; sol/gel and gel glass) to gelation and vitrification phenomena. Much of the information used in the calculation of these curves is based on the torsional braid analysis (TBA), a technique in which a freely oscillating torsional pendulum (TTP), consisting of a glass braid impregnated with the substance to be cured, is used. The freely oscillating torsion

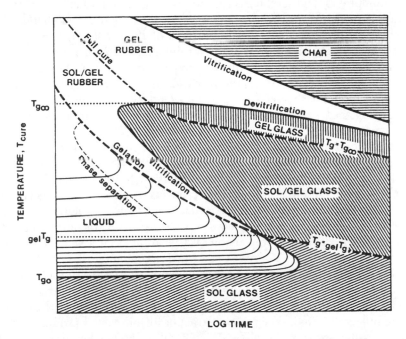

Fig. 6.1. Time–temperature-transformation (TTT) isothermal cure diagram for a solvent-free thermosetting epoxy system. The diagram shows three critical temperatures T_{g0}, gel T_g, $T_{g\infty}$ and the distinct states of matter, i.e. liquid, sol/gel rubber, gel rubber (elastomer), gelled glass, ungelled (or sol) glass and char. The full-cure line, i.e. $T_g = T_g$ (∞), divides the gelled region into two parts: sol/gel glass and fully cured gel glass. Thermal degradation events are responsible for devitrification and the char region. The vitrification process below T_g (gel) has been constructed to be an isoviscous one. The transition region approximates the half width of the glass transition. (By kind permission of GILLHAM 2nd RAPRA Technology Ltd.)

pendulum (TP) can provide qualitative information on the shear moduli (storage and loss) data of solid specimens *versus* temperature or lapsed time. The torsional braid analysis (TBA) uses a composite specimen made simply by impregnating a braid substrate with fluid, which may be a polymer in a volatile solvent or a collection of monomers. This approach provided a significant step forward in the quantification of cure.[11-14] However, since it is based on separating the differences in the damping coefficients of the braid and the fluid, it has been subject to some criticism.

It is assumed that the curing system can be described by a simple reaction mechanism and that no phase separation occurs; then the TTT diagram provides a useful basis for the characterization of curing systems and is primarily a means of defining the various regions and changes in rheological properties. The main features of the TTT diagram can be obtained by measuring the times for various characteristic events to occur during isothermal cure at different temperatures (T_{cure}). These events include the processes of gelation, vitrification, phase separation and devitrification. Phase separation may occur if the system contains a rubber modifying agent, generates a polymer species which becomes insoluble in its own monomer or a crystallizable component. Gelation corresponds to the formation of an incipient infinite molecular network and rheologically leads to long-range elastic behaviour in the thermoset system. Flory[2] has defined the point at which gelation occurs in terms of the degree of conversion of monomer to the final crosslinked matrix. After gelation, the material consists of normally miscible sol (solvent-soluble) and gel (solvent-insoluble) fractions, the ratio of the former to the latter depending upon the degree of conversion. Vitrification occurs when the glass transition temperature, T_g, rises to be equal to or greater than the isothermal temperature of cure. The glass transition corresponds to the point at which cooperative short-scale motion of the crosslinking matrix ceases and vitrification occurs. It is important to appreciate that at this point there may still remain a significant fraction of unreacted groups.[3,6] The system is fluid or rubbery when $T_{cure} > T_g$; it is glassy when $T_{cure} < T_g$. If the T_g decreases through the T_{cure} then the phenomenon of devitrification is observed. At high temperatures, degradation of the matrix initially formed can occur, leading to changes in the structure of the matrix and often a lowering of the glass transition temperature.[15]

The TTT diagram (Fig. 6.1) displays various distinct phases which may be encountered during the cure of a resin. Technologically, the terms A-, B- and C-stage resins are used to designate the sol glass, sol/gel glass and fully cured gel glass, respectively. Figure 6.1 includes a series of zero shear-rate isoviscous contours in the liquid region; successive contours differ by a factor of ten. The vitrification process below T_g (gel) has been constructed to be an isoviscous one. Much of the observed behaviour of a thermosetting system can be explained in terms of the T_g (gel) and T_g (∞) values. As the cure process proceeds at a constant temperature, so the viscosity increases until gelation occurs. The gelation state is generally accepted to correspond to the system achieving a viscosity of the order of 10^4 Pa s. Care must be used in the interpretation of data from the TTT diagram, as it is assumed that the system conforms to isothermal conditions at all times. In practice, the exotherm associated with the curing process usually will occur in the low-viscosity state and hence can be dissipated by the thermal motion of the resin in the mould. However, if the cure is allowed to occur rapidly, the reaction can become a very effective source of heat and the exotherm can raise the temperature many tens of degrees above that chosen for the cure process. This effect will lead to degradation of the resin and the formation of a coloured or charred product.

In general, the cure will proceed until the glass transition of the system equates with that of the cure temperature. Once the system has gelled, then the rate of reaction is suppressed and further reaction proceeds very slowly, indicated by the fact that the system now corresponds to the sol/ gel glass phase. For the system to reach its equilibrium glass transition, T_g (∞), it is usually necessary for the temperature to be elevated to complete the reaction process. If too high a cure temperature is used, then the possibility of degradation becomes important, reflected in either the formation of a structural char or a situation where a thermoset appears to pass through the glassy phase but subsequently returns to the gel rubber phase as a consequence of devitrification. In addition to the above features, it is also possible that phase separation may occur with consequential changes in the rheology in the liquid phase. The large changes in the viscosity parameters during the cure process make it essential that the equipment used is capable of observation over four or five orders of magnitude. The TTT studies of Gillham based on the torsional-braid approach, while applicable to a wide range of systems, are not completely accepted by all rheologists.

In this review, the application of the vibrating-paddle type of instrument will be discussed. Two approaches to the problem will be connected; the vibrating needle instrument VNC promoted by the Rubber and Plastic Research Association (RAPRA) and the vibrating paddle rheometer based on a design developed at Strathclyde University and commercialized by Polymer Laboratories.[16] Both systems have been developed to provide industry with a relatively simple and widely applicable method for quantification of the cure process. One of the prime prerequisites for such an instrument is that it should have a large dynamic range, be robust and easily used. Both instruments can be used with attachments which allow isothermal conditions to be achieved for the cure of the material or can have the probe elements inserted into a reaction medium, allowing the instrument to be used as a process control monitor.

6.2.2. Methods Available for Cure Monitoring

The problem of characterization of the state and progress of a thermoset reaction is one which has been addressed many times.[6,8-12] A wide variety of methods exist for the characterization of cure and include Fourier Transform Infrared spectroscopy {FTIR}; Differential Scanning Calorimetry {DSC}; electrical resistivity; dynamic dielectric permittivity analysis; and rheological techniques. The FTIR and DSC measurements allow direct quantitative data to be obtained on the progress of reaction, and hence allow the degree of conversion to be measured directly. Many thermoset reactions involve the reaction of two groups to form a single stable linkage, as in the case of an epoxy reaction with an amine or a urethane with an amine or diol. The loss of the infrared absorption of a characteristic band can be used quantitatively to follow the progress of the reaction. In the case of epoxy resins, the ring vibrations at approximately $918 \, \text{cm}^{-1}$ is often used and for the urethanes the isocyanate band at approximately $2100 \, \text{cm}^{-1}$ is monitored. Use of this technique allows the effects of vitrification to be quantified and indicates that often cage effects will lead to limitation of the degree of conversion which is achieved. FTIR measurements can also be used to indicate that if a temperature used in cure is subsequently raised, the increased mobility of the reactive species allows the crosslink reactions to proceed. DSC measurements play an important role in monitoring the extent to which reaction occurs at a particular cure temperature and measurement of the residual reaction exotherms can provide quantitative estimates of the degree of conversion. These measurements also have the added advantage of allowing determination of the glass transition temperature and its shift with advancement of cure. Other methods which have been used to monitor cure include measurement of the electrical resistivity of the sample. In this case the processes of gelation and vitrification are marked by a decrease in the conductivity of the matrix. Recently, real time dielectric measurements have provided a more quantitative investigation of the electrical properties of a curing matrix. A number of features can be observed and these include changes in the conductivity with cure; shift of the dipolar relaxation process in the frequency plane as a function of cure; and also changes in low-frequency dielectric loss peaks which can be related to the gelation process. However, the curing process is best defined and determined in terms of changes in mechanical properties, since these are directly related to the generation of the three-dimensional network structure in the system. Torsional braid and related measurements have been found to be very attractive in this context; however, they require the use of specific geometries for this type of investigation. In this article, the use of vibrating-probe methods are considered as an alternative approach to obtaining rheological data.

6.3. VIBRATING NEEDLE CUREMETER (VNC)[8,17]

This method is based around the concept that the damping of a vibrating needle may be related to the process of cure in a reactive system. The heart of the VNC is the vibrator head which contains a moving-coil assembly located via a flexible mounting close to one pole of a permanent magnet (Fig. 6.2). The coil is energized by an AC current to produce a vibration of the same frequency as the oscillating current, which is derived from the amplified output of an oscillator ($10-10^6$ Hz). Typical output voltages from the oscillator (input voltage to the vibrator) are in the range 1–1·5 V and the frequency of operation lies between 35 and 300 Hz. The choice of operating frequency is critical to this type of instrument. With input voltages to the vibrator in the range specified above, the amplitude of vibration at resonance in air will be around 0·1 mm. The low amplitude of vibration allows it to be used as a process cure monitor. With a stable input signal, the peak amplitude of the vibration remains constant and thus any resistance to the movement results in a reduced amplitude of vibration. This amplitude attenuation provides the mechanism for monitoring the progress of cure. A reduction in the amplitude of vibration leads to a detectable change in the back EMF in the electromagnetic coil. This is the output signal from the vibrator. The standard VNC configuration uses a 30 Ω vibrator with vibrating mass (moving coil plus needle) of 7·5 g. When the input voltage is in the region 1–1·5 V, the output signal is around 100 mV for vibrations at 40 Hz in air. Setting up the instrument to produce this output signal under these conditions provides a convenient means of achieving reproducible performance from the instrument.

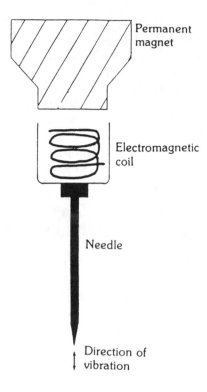

Fig. 6.2. Schematic diagram for the vibrator and needle of the VNC. (By kind permission of RAPRA Technology Ltd.)

6.3.1. Amplitude Attenuation for the VNC[8]

The VNC is usually used close to its resonance condition, and the simplest model for the system is based on the concept of a mass suspended on a spring with one end fixed (Fig. 6.3a). If the spring has a stiffness constant of k_0, then the equation of motion of the system has the form:

$$m\, d^2y/dt^2 + k_0 y = 0 \qquad (6.1)$$

or

$$d^2y/dt^2 + \omega_0^2 y = 0 \qquad (6.2)$$

Fig. 6.3. Spring and dashpot models for the VNC instrument. (By kind permission of RAPRA Technology Ltd.)

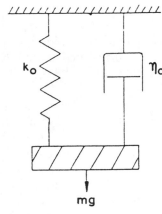

a.

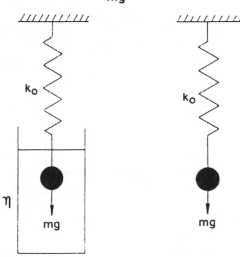

b.

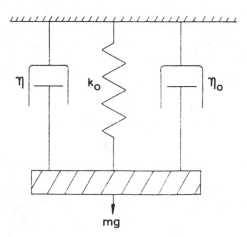

c.

where m is the mass involved in the vibration and ω_0 is the natural frequency of oscillation in radians per second of the oscillator and $\omega_0^2 = k_0/m$.

$$y = y_0 \sin(\omega_0 t) \tag{6.3}$$

In free vibration in air the system will have a natural frequency ω_0 of $\sqrt{k_0/m}$. However, in forced vibration the system is capable of oscillating at any frequency dictated by the driving frequency. When the driving force equals the natural frequency then the resonance condition will be achieved. This effect is readily seen with the VNC when operated in air at approximately 45 Hz (Fig. 6.4).

In practice, there will be some internal friction in the vibrator and a damping term should be included in eqns (6.1) and (6.2). The system is then represented by a spring and dashpot in parallel, as in Fig. 6.3(b). When the needle is placed in viscous media, the model needs to be extended to include additional damping terms (Fig. 6.3(c)). In practice, the sample being monitored will be in a container or attached to a substrate of sufficient mass that a condition can be achieved in which the needle becomes attached to a fixed plane by a flexible element (Fig. 6.3(d)). In this situation the top and bottom of the system are both fixed in space with the central part free to vibrate, and has the form of a spring and dashpot in parallel. The effect of the damping is to reduce the amplitude of the vibration and depends upon the viscosity of the fluid (Fig. 6.4). The amplitude at resonance, as measured by the back emf, is significantly reduced as the viscosity increases to around 200 Pa s in the case of the cane sugar melt.

Interestingly, the position of the resonance peak with respect to frequency changes very little, the top three traces all having their maximum at 43–45 Hz. This is in accord with the simple model presented above; the resonance frequency will shift to lower frequencies as the viscosity is increased.[18,19] In the case of cane sugar, the maximum is broad but has shifted to around 40 Hz.

6.3.2. Output Voltage and Viscosity

The equation of motion of the forced vibration for the system in Fig. 6.3(d) is

$$m\,\mathrm{d}^2y/\mathrm{d}t^2 + (C\eta + \varphi)\,\mathrm{d}y/\mathrm{d}t + k_0 y = f_0 \cos(\omega t) \tag{6.4}$$

where η is the viscosity of the fluid, C is a shape factor for the probe in the curing fluid, φ is the damping term for the internal friction in

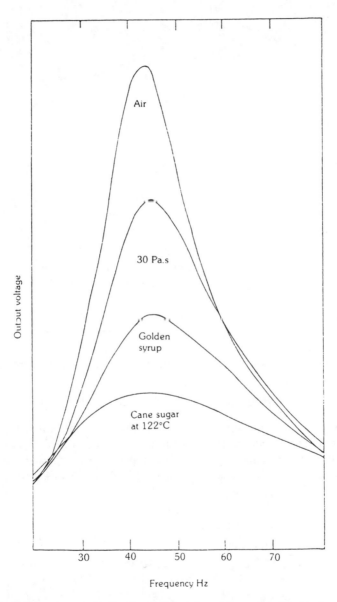

Frequency Hz

Fig. 6.4. VNC resonance peaks for different damping media. (By kind permission of RAPRA Technology Ltd.)

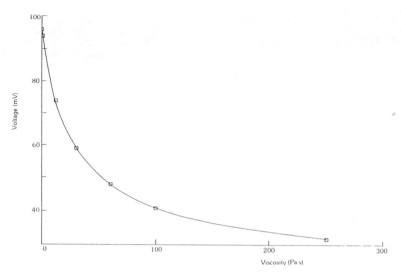

Fig. 6.5. Response of the VNC to silicone fluids of different viscosities. Comparison of experimental data (points) with the plot from a curve-fitting model. (By kind permission of RAPRA Technology Ltd.)

the vibrator and $f_0 \cos(\omega t)$ is the imposed periodic force of angular frequency ω.

The solution to eqn (6.5) for the amplitude at resonance in terms of the output voltage is

$$a_4 \eta^4 + a_3 \eta^3 + a_2 \eta^2 + a_1 \eta + a_0 = 1/V^2 \qquad (6.5)$$

where V is the output voltage and a_4–a_0 are constants for a particular instrument. This model is an oversimplification of the real situation and the parameters obtained should not be considered to have any physical significance. However, the above relationship does provide a remarkably

TABLE 6.1
Values of the Five Constants for Eqn (6.5)

Constants	Values
a_4	$2 \cdot 173 \times 10^{-14}$
a_3	$1 \cdot 927 \times 10^{-11}$
a_2	$-1 \cdot 821 \times 10^{-8}$
a_1	$6 \cdot 489 \times 10^{-6}$
a_0	$1 \cdot 065 \times 10^{-4}$

good fit with experimental data (Fig. 6.5). The plot is obtained by graphing the VNC output voltage from a series of liquid silicones of different viscosities, the solid line representing the theoretical fit of the data using eqn (6.5), with values of the constants presented in Table 6.1.

6.3.3. Application of the VNC to Monitoring Viscous Changes

6.3.3.1. Determination of Pour Time, Pot Life or Application Time
In this mode, the VNC is set up in its resonance mode in air and the output amplitude is monitored as a function of time (Fig. 6.6), and the corresponding viscosities can be calculated using Fig. 6.5. For a silicone RTV formulation of initial viscosity 60 Pa s, which doubles its viscosity on cure, the VNC output would be expected to drop from 48 to 39 µV, a decrease which is readily detectable. It should, however, be recognized that measurement of the absolute viscosity is not a requirement for the effective use of the VNC in monitoring these early stages of cure. The limits of flow for effective transfer of material are dependent entirely upon the processing step being undertaken. Filling a complex mould or wetting and impregnating a reinforcing matrix may require a lower viscosity than for other operations, and the ability to monitor the process directly and continuously is the ideal situation. It is possible with practice to read

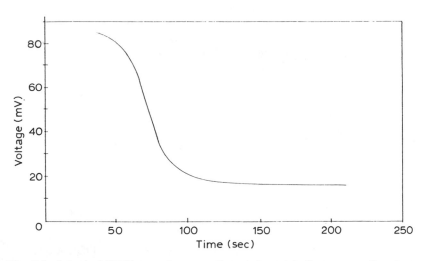

Fig. 6.6. A typical VNC trace for operation at the original resonance frequency. (By kind permission of RAPRA Technology Ltd.)

R. A. Pethrick

directly from the amplitude trace the initial stages of cure and it is not essential that the viscosities are computed.

6.3.3.2. *Delayed-Action Catalysis*

One method of enhancing control in the early stages of cure is by the use of a delayed-action catalyst.[20] In most practical situations it is essential that this enhanced control is achieved without the penalty of changing the final cure rates. An example of such a system is the cure of polyurethanes using an organomercurial catalyst[21] (Fig. 6.7). The traces were obtained at 40 Hz and correspond to the cure of two polyether/polymeric MDI cures, one catalysed by an amine and the other catalysed by an organomercury salt. The difference in the early stages of cure is apparent even though these two mixes had been formulated to give the same gel time of 8 min.

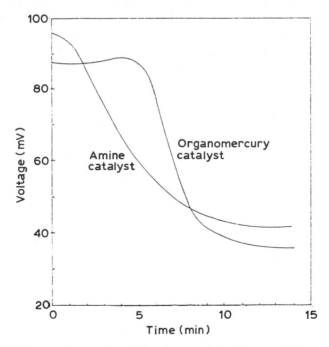

Fig. 6.7. VNC trace for a polyurethane cure catalysed by two different catalysts. (By kind permission of RAPRA Technology Ltd.)

6.3.3.3. *Application of the VNC to the recognition of gelation characteristics*

The above example illustrates that the use of gel-time determination alone may be of little value in distinguishing key differences in processibility of a mix. However, as was discussed in the earlier sections, the gel time is an important parameter in its own right. Using the VNC it is possible to observe the shift in the resonance frequency with cure time (Fig. 6.8). As the polyurethane cures, so the resonance frequency shifts to a higher value, changing from a frequency of approximately 40 Hz at 2 min to a value of 200 Hz after 1250 min. The change in resonance frequency can be modelled using an extension of the simple models discussed previously. In the simplest case, the undamped spring can be modified by attachment of a second spring (Fig. 6.9). The equation of motion of the mass m is given by

$$m \, d^2y/dt^2 + (k_0 + k_1)y = 0 \tag{6.6}$$

$$(k_0 + k_1)/m = \omega_0^2 \tag{6.7}$$

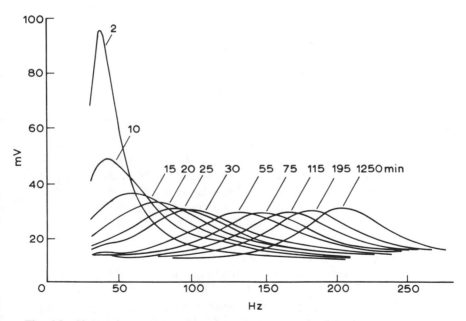

Fig. 6.8. Change in resonance frequency during the cure of a polyurethane. (By kind permission of RAPRA Technology Ltd.)

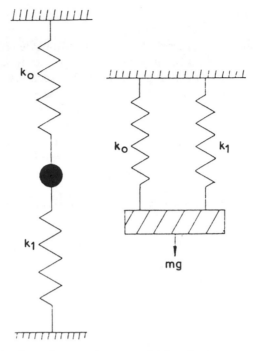

Fig. 6.9. Combination of two springs—model for the latter stages of cure. (By kind permission of RAPRA Technology Ltd.)

The addition of a second elastic element increases the natural vibrational frequency and consequently the frequency of resonance for forced vibrations. Since a shift to higher frequency has not been associated with viscous changes, it may be inferred that the observation of such a shift is diagnostic of additional elasticity in the vibrating system.

6.3.4. Gel Point of a Polyurethane

The gel point is defined as the condition at which an infinite network is first formed. Flory[2] has described the phenomenon in some detail and has used a statistical approach to predict the onset of gelation; this topic has been discussed in detail by Durand.[6] The critical conversion before gelation can occur is defined for a step-growth cure involving difunctional co-reagents (e.g. A–A and B–B) and a third reagent C polyfunctional in A, by[22]

$$P_{crit} = [r + r\rho(f - 2)]^{\frac{1}{2}} \qquad (6.8)$$

where r is equal to unity for a stoichiometric cure, f is the functionality of the polyfunctional reagent and $\rho = N_c f/(2N_A + N_C f)$, with N_A and N_C the number of moles of A–A and C, respectively. The conversion p at any time can be estimated from the initial concentration C_0 of either functional group in a stoichiometric cure, according to

$$C_0 k t = p/(1 - p) \qquad (6.9)$$

where k is a second-order rate constant. A plot of the data obtained for a polyurethane is shown in Fig. 6.10. A striking feature of this plot is the initial plateau, corresponding to a constant resonance frequency, which continues until a definite stage in the cure when this frequency starts to climb. The point when this resonance frequency starts to shift is conveniently determined from the graph, and corresponds in this case to a time of 36 min. Studies by RAPRA have shown that using appropriate numbers for the parameters in eqns (6.9) and (6.10) it can be shown that the critical conversion for gelation is 75%. Close correspondence has been shown between the values estimated using the VNC and those obtained from spectroscopic measurements.

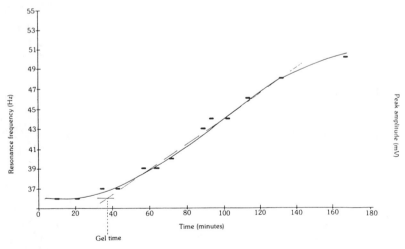

Fig. 6.10. Change in resonance peak frequency with time during polyester–polyurethane cure at 50°C. (By kind permission of RAPRA Technology Ltd.)

6.3.4.1. Work-Life Determination from Gelation Behaviour
The gel point is the point beyond which cure mix cannot be caused to flow; this has implications for the *working time*, a period leading up to this irreversible change. Whilst the length of this period may be somewhat process-dependent, some clues might be gained from the plot of resonance shifts (Fig. 6.10). Although a discrete point has been determined for gelation by extrapolation, there is evidence of some movement of the resonance frequency prior to this time, in this case at approximately 20 min, and represents the upper limit for no detectable resonance shift, related to *work life* or *working time*.

6.3.5. Post Gelation Studies with the VNC
The correlation between the development of elasticity and the shift in resonance frequency provides in principle a simple approach for the determination of cure time. This would involve monitoring the resonance frequency with time (Fig. 6.10), until a steady state is achieved once more at the final resonance frequency.

Whilst this approach is certainly capable of being followed, in practice useful information on the development of cure may be obtained by an alternative method which is simpler to operate and involves monitoring the voltage time change at a fixed frequency. An example of this approach

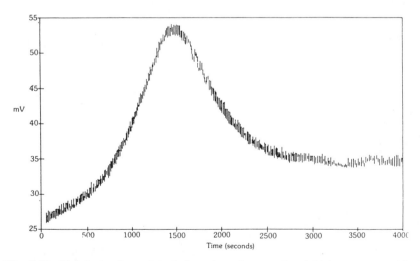

Fig. 6.11. Determination of tack-free time for a polysulphide cure. (By kind permission of RAPRA Technology Ltd.)

is presented in Fig. 6.11. A polysulphide sealant was cured via a thiol oxidation reaction base catalysed with manganese dioxide (100 parts MnO_2, 10 parts of 4-ethylmorpholine). The final cure gives a resonance peak with the VNC at around 200 Hz. The tack-free stage is achieved before cure and a comparison with the ASTM method showed that this was obtained after around 25 min cure when the VNC-determined resonance frequency was 160 Hz. Monitoring the cure with the VNC operating continuously at 160 Hz gives a maximum when this condition of resonance is achieved (Fig. 6.11), which perhaps not surprisingly occurs after 25 min. Thus the trace in Fig. 6.11 gives a maximum at the tack-free time. The generalization from this is that the VNC has been tuned to provide the time when a predetermined level of cure has been achieved. If this corresponds to a tack-free time or demould time for a cured network derived from a given class of thermosets, then the method can be applied to related formulations.

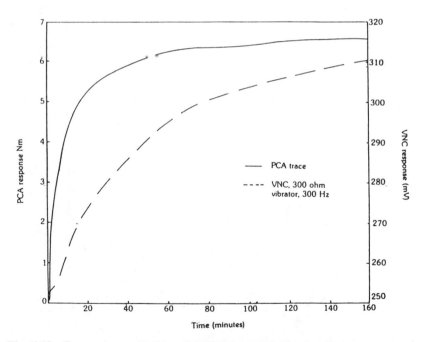

Fig. 6.12. Comparison of PCA and VNC data (300 Hz) for the later stages of a PU cure. (By kind permission of RAPRA Technology Ltd.)

6.3.6. Determination of Cure Time

One convenient method for the determination of the cure time is to estimate it from the voltage/time change at a constant frequency, which is
at least as high as that required to give resonance in the final cured state.
In practice, this is likely to involve monitoring at a frequency in excess
of 200 Hz, when the output trace may sometimes be a little noisy. If this
is the case, then useful improvements can be obtained by switching to a
higher impedance vibrator. The cure trace in Fig. 6.12 is that obtained at
300 Hz when the normal 30 Ω vibrator was replaced by a 300 Ω vibrator.
The trace shows that the VNC is capable of responding to changes occurring after 2 h in a polyurethane system which gelled after 2 min. In this
mode, the VNC is operating in the region where traditionally shear-
modulus devices have been used for cure monitoring. One such device is
the Wallace–Shawbury Precision Cure Analyser (PCA),[23] which shows
similar profiles to the VNC.

Comparison of the response of these devices (Fig. 6.13), indicates that
whilst the PCA is useful for the detection of the latter stages of cure it is
not capable of being used to monitor the earlier stages of cure, a time
window to which the VNC is ideally suited as is the Strathclyde
Curemeter.

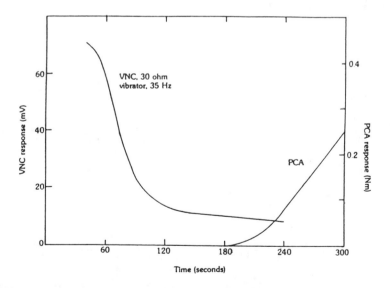

Fig. 6.13. Comparison of PCA and VNC data (35 Hz) for the early stages of a
PU cure. (By kind permission of RAPRA Technology Ltd.)

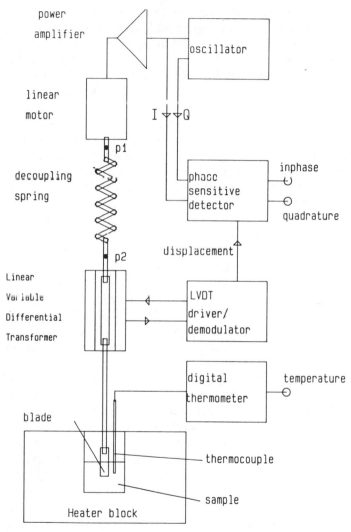

Fig. 6.14. Schematic diagram of the Strathclyde curemeter. (By kind permission of RAPRA Technology Ltd.)

6.4. STRATCHLYDE CUREMETER[24-26]

The discussion of the VNC indicates that operation of resonance of the needle curemeter can lead to significant loss of sensitivity, and whilst the VNC has great versatility, it is limited in measurement of very large ranges of modulus change. The *Strathclyde Curemeter* (SC) was developed independently, but has certain features in common with the VNC. The method of probing the system is a paddle rather than a needle; however the dimensions of the probe are very small, making the difference a little academic. The other common feature is that the instrument involves the damping of a vibration by the curing process. The apparatus (Fig. 6.14), is similar in format, having a simple vibrating paddle as the sensor, extending from a housing in which the sensor, driver and spring element are housed. The paddle is connected to a constant amplitude and frequency linear motor by a spring. The net result is that the amplitude and phase of the motion of the bottom of the spring (p2) are directly related to the degree of damping produced by the gelling fluid. It is easy to see that the above configuration is directly analogous to the VNC with the addition of the spring element. In principle, the SC can do all that the VNC can, but has the additional advantage of enhanced sensitivity accompanying the use of phase-sensitive detection. The motion of the probe is sensed by a linear variable differentiable transformer (LVDT). Comparison of the in-phase components of the amplitude of the motion and phase shift can be used to calculate the rheological properties of the curing system.

The limitations of the method are associated with the structure of the sensor system. A large plate would be difficult to support and, if placed in a narrow gap, would produce excessive stirring of the fluid. The size and shape of the probe have been chosen so as to provide the largest possible surface area whilst also producing a minimal disturbance of the fluid. During the development of this method a wide variety of probe structures was tried, ranging from flat discs to cylinders, large plates and rings. The structure used appears to give the correct compromise between sensitivity and the minimal perturbation of the system, whilst allowing large viscosity changes to be measured during the cure process.

The dynamic range of the curemeter is defined by the spring and the frequency of operation, and for the current system it is possible to explore the range $10-10^5$ Pa s. Use of a weaker or stronger spring in principle allows a lower or higher range to be investigated. The demodulated output of the LVDT is compared with the driving force of the linear motor (p1) using a phase-sensitive detector. Both the in-phase and quadrature components of the probe displacement are recorded and stored in a computer. A measurement of the temperature in the reaction vessel is made via a thermocouple which is placed in the stainless steel sheath close to the probe. The temperature of the sample itself is controlled via a temperature block, but the head can be detached to provide a sensor system to be used in the equivalent mode as the VNC. A range of frequencies for the driver was explored and a value of 2 Hz was chosen to give an optimum for most systems. The apparatus has a natural resonant frequency which is defined by the spring constant, and the mass of the probe and rod. In the current system, this has a value of about 60 Hz, well away from the measurement frequency used in the current system. The amplitude of the probe was also varied between 0·1 and 3 mm, a value of 0·5 mm appeared to be optimal for many systems and the area of the plate contracting the fluid was typically 3 mm^2.

A mathematical analysis of the system shown in Fig. 6.14 is based on the following assumptions:

  (i) The amplitude of the motion of the spring (p1) is fixed, irrespective of the probe amplitude. In practice, the difference in compliance between the motor suspension and the coupling spring means that the motor is practically independent of the probe amplitude. The motor suspension has an effective compliance which is a factor of one hundred greater than that of the spring.

  (ii) The mass of the probe is negligible, i.e. resonant behavior of the probe and spring can be neglected. Since the operating frequency is more than a decade lower than the resonant frequency of the system, this assumption is justified.

  (iii) The change in the probe/material contact area with the probe movement is negligible. Some finite change in the area is inevitable, but with the correct design of the probe the effects are minimal and would in any case show up as 2nd harmonic components in the probe displacement. Such harmonics are rejected by the phase-sensitive detector.

  (iv) The force of the material on the probe is due purely to the viscosity of the fluid. Stirring of the liquid is considered negligible.

Equating the forces at the lower end of the coupling spring and neglecting the mass effects of the spring and probe gives

$$k(p_1(t) - p_2(t)) + \eta^* C \, \mathrm{d}p_2/\mathrm{d}t = 0 \tag{6.10}$$

where k is the spring constant, p_1 and p_2 are the instantaneous displacements from equilibrium at points p1 and p2, respectively, η^* is the complex viscosity of the material and C is a geometric factor related to the probe/material contact area.

Since $p_1(t)$ is a sinusoidal function, the differential equation can be written in complex notation as

$$k(p_1 - p_2) + j\eta\omega C p_2 = 0 \qquad (6.11)$$

where the values of p_1, p_2 and η are also dependent upon the frequency ω. Equating real and imaginary components, assuming that η is real gives

$$\dot{p}_1 - \dot{p}_2 - (j\eta\omega C/k)\ddot{p}_2 = 0 \qquad (6.12)$$

and

$$\ddot{p}_2 - (\eta\omega C/k)\dot{p}_2 = 0 \qquad (6.13)$$

where $\dot{p}_2$ and $\ddot{p}_2$ are the real and imaginary components of p_2. Solving for $\dot{p}_2$ and $\ddot{p}_2$ gives

$$\dot{p}_2 = p_1/(1 + (\eta\omega C/k)^2) \qquad (6.14)$$

and

$$\ddot{p}_2 = p_1\eta\omega C/k/\{1 + (\eta\omega C/k)^2\} \qquad (6.15)$$

The viscosity η can therefore be obtained directly from observation of the movement occurring at point p2 relative to that at point p1. In the current system, the linear motor is sufficiently strong for the amplitude of the motion at point p1 to be assumed constant during the whole of the experiment.

6.4.1. Calibration of Apparatus

In order to test the validity of this method, the damping coefficients for a liquid whose viscosity has been accurately determined as a function of temperature were obtained. The liquid used is an oligomer of polyphenyleneoxide (Santovac 5) and is capable of being supercooled to a highly viscous state. The sample used in this study had been previously measured by the Glasgow University viscoelastic group.[27] The variation of the amplitudes as a function of the viscosity predicted by the theory presented above (solid and dashed lines, Fig. 6.15), fit the experimental data very well. The theoretical data are calculated from the expression for the real and imaginary components with the value of C adjusted to produce coincidence of the theoretical and experimental maxima in the imaginary

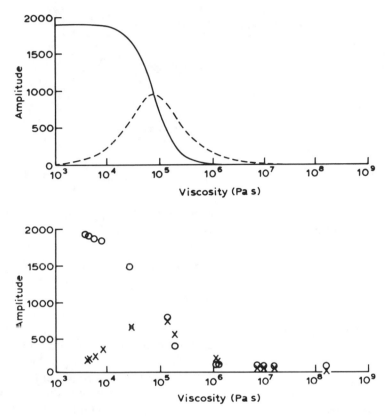

Fig. 6.15. Amplitude components for SANTOVAC 5 at various viscosities. Measured: ○ real, × imaginary. Calculated: real ——, Imaginary – – –. (By kind permission of RAPRA Technology Ltd.)

component. A check of the validity of the theory can be seen from a comparison of data obtained using the SC and Rheometrics (Germany) rheometer (Fig. 6.16). The agreement with simple theory is good over the range of viscosities examined, allowing for the arbitrary geometrical factor *C*. A more rigorous analysis would allow calculation of the latter factor[28] but for most practical purposes use of a calibration approach is preferred. The fact that the amplitude shows the simple sigmoidal behaviour as the viscosity changes implies that we are not looking at a more complicated case of an infinite boundary problem.[28] This is presumably because, at the present frequency and for the viscosities examined, the

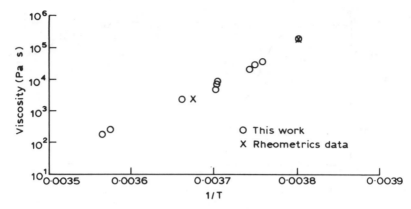

Fig. 6.16. Arrhenius plot for Santovac 5. Key: ○ this work, × Rheometrics data. (By kind permission of RAPRA Technology Ltd.)

smallest wavelengths are much greater than the dimensions of the sample. It is interesting to note that the form of the curves are very similar to those presented for the VNC discussed above.

6.4.2. Cure of an Epoxy Resin System

To illustrate the sort of data which are obtained during a cure process, the curves for the stoichometric cure of the diglycidyl ether of bisphenol A (DGBA) and diamino diphenyl sulphone (DDS) are presented (Fig. 6.17). At low temperatures, gelation of this system and the viscosity increase proceeds very slowly over relatively long periods of time. In contrast, the plots obtained at higher temperatures indicate a much more rapid increase in the viscosity, consistent with the faster rate of reaction and the lower possibility of relaxation of the matrix during the cure process. The sensor monitoring the temperature indicates that the system dramatically exotherms at higher temperatures. The cure profile can give information on a variety of different parameters as discussed previously for the VNC. The point at which the damping is observed to be reduced (t_1 in Fig. 6.17) is indicative of the *pot life* of the resin at 120°C. Beyond this point, the resin begins to build a structure which will be incapable of further flow and produces a high-quality moulding. The point at which

Fig. 6.17. Amplitude components, exotherms and viscosities for the cure of DGBA–DDS at various temperatures. (By kind permission of RAPRA Technology Ltd.)

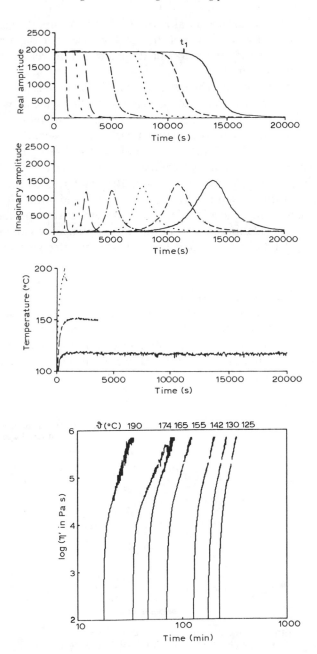

the derived viscosity reaches a value of 10^4 Pa s is normally accepted as the *gelation* point. By analogy with the analysis used for the VNC, it is possible also to identify the peak maximum which correlates with the *working life* of the resin.

Definition of the point at which the system is fully cured is more difficult to define; however, it appears to coincide with the point at which the quadrature component has reached a zero value. The values obtained from this analysis are in good agreement with other observations on this system (Fig. 6.18).[29] In addition, it is possible to observe that, related to this peak, an increase in the temperature is often observed, which corresponds to the exotherming of the matrix material during cure. A large exotherm during cure often leads to microvoid formation and a poor moulding. Hence, the measurement of the temperature profile during the cure is often a valuable piece of information which allows an optimization of the gel time without the disadvantageous effects of microvoid formation.

6.4.3. Cure of a Powder Resin System

Environmentally, the use of powders is favourable to the use of solvent-suspended polymer systems and as a result, powder coatings have become of considerable interest in recent years. In the case of determining the cure of a powder coating, an additional complication is associated with

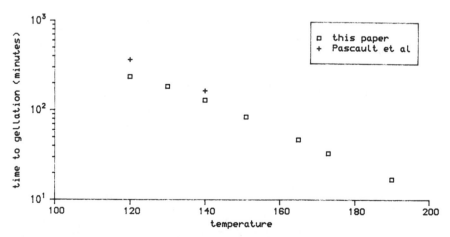

Fig. 6.18. Cure time *versus* cure temperature for DGBA–DDS. Key: □ this work, + data of Grillett *et al.*[29] (By kind permission of RAPRA Technology Ltd.)

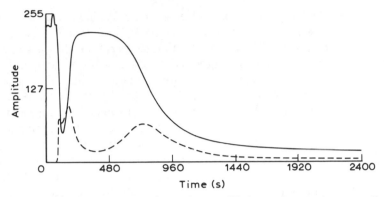

Fig. 6.19. The curing process of a polyester-filled powder coating—amplitude components. Key. ——— real, - - - imaginary. (By kind permission of RAPRA Technology Ltd.)

the melting of the powder to form a coherent film. In Fig. 6.19 a curve is presented for a cure cycle for a polyester-filled powder coating. The initial stages of the curve reflect the changes of the powder during its melting and thinning stages. It can be seen that the powder will initially wet-out the blade, marked by the increase in the quadrature component and, since the resin is at a temperature lower than that used for cure, will have a greater viscosity. As it heats up to the curing temperature, so the viscosity will reduce and this is reflected by the real amplitude increasing and the quadrature decreasing. The now-fluid powder coating undergoes a cure profile which is similar to that observed in the fluid epoxy resin system. In practice, it is found that a useful characteristic of the curing process is to take the time difference between the melt peak and that associated with the gelation process. These measurements have been found to be a very sensitive test of the effects of the interaction between filler and resin.

6.4.4. Plastisol Systems

The conversion of a dispersion of a solid polymer into a plasticized solid is technologically very important and forms the basis of the *plastisol* process. The plastisol usually consists of a dispersion of polyvinylchloride (PVC) particles in a poor solvent: an alkylphthalate. On heating, the particles of PVC swell and expand to form an entangled gelled matrix structure. At low temperature (Fig. 6.20) the rate of swell is slow and this is indicated by the slow monotonic increase in the viscosity as the size of the particles dispersed in the plasticizer increase. At higher temperature, the rate of swelling is dramatically increased, and this is once more reflected in the observed rheology. At high temperature, the curves indicate that a limiting viscosity is attained, whereas at lower temperatures the viscosity appears to increase continuously to some high value. At high temperatures, the resultant matrix will have a glass transition temperature which is below that used for the generation of the plastisol, and a flexible product is obtained. At low temperature the glass transition temperature of the matrix is above that of the processing temperature and a rigid matrix is obtained. The curves therefore not only give information on the rate of gelation of the plastisol, but also indicate the nature of the final product. These observations are analogous with those presented for the VNC when operated at high frequency.

6.5. CONCLUSIONS

In this short review, the utility of the needle rheometer has been presented. Both the VNC and SC are capable of determining a number of very important characteristics of a curing system. The SC has the advantage of sensitivity and avoids the necessity of using variable frequency, although like the VNC it is available as an additional variable if required. It is possible from the observed rheological curves to define technologically important variables, such as pot life, cure time, etc., and to provide a tool for the quality control of reactive systems.

ACKNOWLEDGEMENTS

The author wishes to thank Prof. K. J. Gillham and Dr K. W. Scott for permission to use diagrams and data from their research as part of this review. I would also wish to thank Dr F. S. Baker, Carter Baker Enterprises Ltd, 2 Marshbrook Business Park, Marshbrook, Church Stretton, Shropshire, UK, SY6 6QE for provision of the Strathclyde Curemeter used in these studies and with help in design of the

---

Fig. 6.20. Amplitude components and viscosity of plastisols cured at various temperatures. Key: —— 78°C, · · · · · 94°C, – – – – 105°C, ——— 122°C. (By kind permission of RAPRA Technology Ltd.)

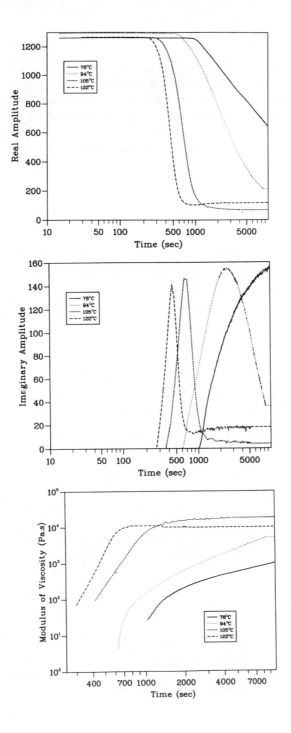

original instrument. The Strathclyde Curometer has recently been redesigned and is available commercially from Polymer Laboratories plc, Loughborough. It is designated the Thermals Scanning Rheometer (TSR) and incorporates thermal scanning facilities in addition to the isothermal capabilities described in this article. The support of the SERC, Tioxide, ICI and my various students and coworkers, in particular Dr D. Hayward, are gratefully acknowledged.

REFERENCES

1. R. W. Whorlow, *Rheological Techniques*, Ellis Horwood, Chichester, 1980.
2. P. J. Flory, *Chem. Rev.*, 1946, **39**, 137.
3. M. Gordon, *High Polymers*, Iliffe Books, London, 1963, p. 8.
4. C. W. Macosko, *Brit. Polym. J.*, 1985, **17**, 239.
5. F. G. Musatti and C. W. Macosko, *Polym. Eng. Sci.*, 1973, **13**, 235.
6. D. Durand, *Polymer Yearbook 3*, R. A. Pethrick (Ed.), Harwood Academic Publishers, London, 1984, p. 229.
7. B. Erman, *Crosslinking and Scission in Polymers*, O. Gurven (Ed.), Nato ASI Series, Kluwer Academic Publishers, Dordrecht, 1988, **292**, 153.
8. B. G. Willoughby, K. W. Scott and D. Hands, *Flow and Cure of Polymers— Measurement and Control*, RAPRA Technology Ltd, Shrewsbury, 1990.
9. G. S. Springer, in *Progress in Science and Engineering of Composites*, T. Hayashi, K. Kawaka and S. Umekawa (Eds.), Japan Society for Composite Materials, Tokyo, 1983, p. 23.
10. M. D. Dusi, W. I. Lee, P. R. Ciriscioli and G. S. Springer, *J. Composite Materials*, 1987, **21**, 243.
11. J. K. Gillham, *Flow and Cure of Polymers—Measurement and Control*, RAPRA Technology Ltd., Shrewsbury, 1990.
12. J. K. Gillham, *Developments in Polymer Characterization*, J. V. Damkins (Ed.), Vol. 3, Applied Science Publishers Ltd., London, 1982, Ch. 5, p. 159.
13. J. K. Gillham, *Brit. Polym. J.*, 1985, **17**, 224.
14. J. K. Gillham, *Crosslinking and Scission in Polymers*, O. Gurven (Ed.), Nato ASI Series C, Kluwer Academic Publishers, Dordrecht, 1990, **292**, 171.
15. R. A. Pethrick and I. D. Maxwell, *Polym. Deg. and Stab.*, 1983, **5**, 275.
16. Polymer Laboratories, Thermal Sciences Division, The Technology Centre, Epinal Way, Loughborough, UK, LE11 0QE.
17. RAPRA Technology Ltd., Shawbury, Shrewsbury, Shropshire, UK, SY4 4NR.
18. R. B. Morrison, *Concise Physics*, Arnold, London, 1962, p. 15.
19. J. P. den Hartog, *McGraw-Hill Encyclopedia of Science and Technology*, Vol. 8, McGraw-Hill, New York, 1971, p. 247.
20. D. Grief, N. Mars, N. Burt, B. G. Willoughby and K. Scott, Studies on an energetically plasticized polyurethane binder. Paper presented at ADPA Symposium, New Orleans, 18–20 April, 1988.

21. K. W. Scott, The vibrating needle curemeter—Application to polyurethanes. Paper presented at PMA Fall Meeting, Itasca IL, 22–25 October, 1989.
22. S. H. Pinner, *J. Polym. Sci.*, 1956, **21**, 153.
23. D. Hands, R. H. Norman and P. Stevens, A new cure meter. Paper presented at PRI Rubber Conference 84, Birmingham, 12–15 March, 1984.
24. S. Affrossman, A. Collins, D. Hayward, E. Trottier and R. A. Pethrick, *J. Oil and Colour Chemists Ass.*, 1989, 452.
25. S. Affrossman, D. Hayward, A. McKee, A. MacKinnon, D. Lairez, R. A. Pethrick, A. Vatalis, F. S. Baker and R. E. Carter, *Rheology of Food, Pharmaceutical and Biological Materials with General Rheology*. R. E. Carter (Ed.), Elsevier Applied Science, London, 1989, p. 304.
26. D. Hayward, A. MacKinnon, R. A. Pethrick, F. S. Baker, J. Ferguson, R. E. Carter and J. H. Daly, *Third European Rheology Conference*, D. R. Oliver (Ed.), Elsevier Applied Science, London, 1990, p. 211.
27. A. J. Barlow, A. Eirginsav and J. Lamb, *Proc. Roy. Soc. A*, 1969, **309**, 469.
28. J. D. Ferry, *Viscoelastic Properties of Polymers*, Wiley, New York, 1980, p. 116.
29. A. C. Grillett, J. Galy and J. P. Pascault, *Polymer* (to be published).

Chapter 7

Dynamic Mechanical Analysis Using Complex Waveforms

B. I. NELSON AND J. M. DEALY

Department of Chemical Engineering, McGill University, Montreal, Quebec, Canada

7.1. INTRODUCTION

Small-amplitude oscillatory shear is a widely used method of determining the linear viscoelastic properties of materials. In the classical version of the technique, the material is subjected to a sinusoidal shear strain of amplitude γ_0 and frequency ω, such that the shear strain as a function of time is

$$\gamma(t) = \gamma_0 \sin(\omega t) \qquad (7.1)$$

If the response is linear, i.e. if the strain amplitude is sufficiently small, the resulting shear stress will also be sinusoidal:

$$\sigma(t) = \sigma_0 \sin(\omega t + \delta) \qquad (7.2)$$

where δ is the phase angle or mechanical loss angle and σ_0 is the stress amplitude. Furthermore, σ_0 at a given frequency is proportional to γ_0, again if the strain is sufficiently small that the response is linear.

Linear viscoelastic behaviour can be described in terms of the amplitude ratio or 'dynamic modulus', $G_d \equiv \sigma_0 / \gamma_0$, and the phase angle, δ, both as functions of frequency. It is customary to rewrite eqn (7.2) using a trigonometric identity as follows:

$$\sigma(t) = \gamma_0 [G' \sin(\omega t) + G'' \cos(\omega t)] \qquad (7.3)$$

where $G'(\omega)$ is the storage modulus and $G''(\omega)$ is the loss modulus, both of which are functions of frequency. The values of these functions at a given frequency can be calculated from the amplitude ratio and phase angle using

$$G' = G_d \cos(\delta)$$
$$G'' = G_d \sin(\delta) \qquad (7.4)$$

where G_d is the dynamic modulus.

The complex modulus, $G^*(\omega)$, is a complex-valued function whose real and imaginary parts are the storage and loss moduli, respectively.

$$G^*(\omega) = G'(\omega) + jG''(\omega) \qquad (7.5)$$

where $j = \sqrt{-1}$.

The dynamic modulus, G_d, is equal to the magnitude of the complex modulus, $|G^*|$. $G^*(\omega)$ can be calculated directly from experimental data, as is explained in Section 7.2.

The storage and loss moduli, if measured over a wide range of frequencies, provide a detailed picture of the viscoelastic nature of the material in its undeformed state. They are widely used in the characterization of polymeric liquids, and empirical methods have been proposed for relating them to an average molecular weight[1] or to the molecular weight distribution.[2]

The standard method of measuring $G'(\omega)$ and $G''(\omega)$ for a material is to perform a series of sinusoidal shear experiments at various frequencies and to build up a response curve one point at a time. This is not an efficient use of time or information, as one must sample an entire waveform to get a single datum. However, if one uses a complex strain history, with a broad-frequency content, rather than a sinusoidal strain, information about the response over a range of frequencies can be determined in a single measurement. To facilitate this, a discrete Fourier transform (DFT) can be used to analyze the complicated stress signal to yield values of the storage and loss moduli at a number of frequencies. In this way, the shape of the response curves can be established very quickly.

Sometimes the crossover modulus and frequency (where $G' = G''$) are of special interest, and a number of points in the vicinity of the crossover are needed to accurately determine these values. If the material can be probed at several frequencies simultaneously, it may be possible to determine the crossover point using a single waveform. The resulting speed of measurement may be essential if one wishes to track a time-varying phenomenon, such as a curing reaction, or to use the technique as a basis for the operation of a rheometer to be used for process control (see Chapter 10).

In this chapter, we first describe the basic techniques for analyzing data obtained using complex waveforms and then list several waveforms that have been proposed for this type of application. Finally, we present an example of the use of the technique in which the strain signal is a pseudo-random binary sequence.

7.2. FREQUENCY ANALYSIS OF COMPLEX WAVEFORMS

The discrete Fourier transform (DFT) separates a discrete time series into its Fourier components. In other words, it calculates the amplitudes of the sine and cosine waves at each of the discrete frequencies that make up the Fourier-series representation of the discrete signal. For the Fourier series

$$f(t) = \sum_{i=0}^{N-1} a_i \cos(\omega_i t) + b_i \sin(\omega_i t) \qquad (7.6)$$

the DFT yields the coefficients a_i and b_i. The DFT can be used to analyze the stress signals generated by a complex strain history by finding the amplitude of the stress response for any particular frequency component.

To understand how the analysis works, consider a sinusoidal strain with a frequency ω_0. The strain and resulting stress are given by eqns (7.1) and (7.2), which are plotted in Fig. 7.1.

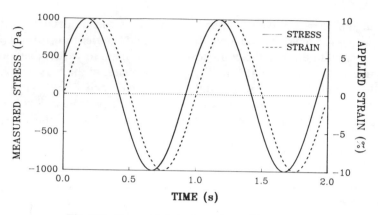

Fig. 7.1. Sinusoidal strain and resulting stress.

As mentioned at the beginning of the chapter, the quantities arising directly from the experimental data are G_d and δ, and we will base our analysis on these. Approaching the analysis from the frequency viewpoint, both the stress and the strain are transformed into the frequency domain.

$$\gamma^*(\omega) = \text{DFT}[\gamma(t)]$$
$$\sigma^*(\omega) = \text{DFT}[\sigma(t)]$$
(7.7)

The transform produces complex-valued functions of frequency that are zero at every frequency except ω_0, where

$$|\sigma^*(\omega_0)| = \tfrac{1}{2}\sigma_0 \qquad |\gamma^*(\omega_0)| = \tfrac{1}{2}\gamma_0$$

(see Section 7.32 for an explanation of the factor $1/2$). The quantities G_d and δ are then calculated as follows:

$$G_d = |G^*(\omega_0)| = \frac{|\sigma^*(\omega_0)|}{|\gamma^*(\omega_0)|}$$
(7.8a)

$$\delta = \text{phase}[\sigma^*(\omega_0)] - \text{phase}[\gamma^*(\omega_0)]$$
(7.8b)

The operations of eqns (7.8a) and (7.8b) are equivalent to the complex division of $\sigma^*(\omega_0)$ by $\gamma^*(\omega_0)$. $G^*(\omega_0)$, containing both amplitude and phase information, can be determined directly from the frequency-domain data; all other functions can then be obtained from $G^*(\omega_0)$:

$$G^*(\omega_0) \equiv \frac{\sigma^*(\omega_0)}{\gamma^*(\omega_0)}$$

$$\delta(\omega_0) = \text{phase}[G^*(\omega_0)] \qquad (7.9)$$

$$G'(\omega_0) = \text{real}[G^*(\omega_0)]$$

$$G''(\omega_0) = \text{imag.}[G^*(\omega_0)]$$

Consider next a strain history composed of two superposed sinusoids having frequencies ω_1 and ω_2 (see Fig. 7.2).

$$\gamma(t) = \gamma_1 \sin(\omega_1 t) + \gamma_2 \sin(\omega_2 t)$$

$$\sigma(t) = \sigma_1 \sin(\omega_1 t + \delta_1) + \sigma_2 \sin(\omega_2 t + \delta_2)$$

One of the basic axioms of linear viscoelasticity is that the stress in the material at any time, t, is the sum of the stresses resulting from each of the individual strains that the material has experienced at past times. This implies that the stress resulting from a composite strain waveform is simply the sum of the stresses arising from the individual components of the strain. In this case, the stress due to the superposed sinusoids is the sum of the responses the individual sinusoids would have caused alone. Since a time-based method of analysis would be very difficult to use, a frequency-based analysis is highly advantageous.

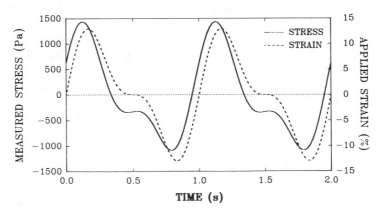

Fig. 7.2. Strain composed of two superposed sine waves and resulting stress response.

Now there are two frequencies, ω_1 and ω_2, at which the transformed strain and stress functions have non-zero magnitudes. The analysis performed at ω_0 for the single sinusoid can now be performed at ω_1 and ω_2 for the multiple sinusoid. The stress and strain signals are transformed into the frequency domain, giving:

$$|\gamma^*(\omega_1)| = \tfrac{1}{2}\gamma_1 \qquad |\gamma^*(\omega_2)| = \tfrac{1}{2}\gamma_2$$

$$|\sigma^*(\omega_1)| = \tfrac{1}{2}\sigma_1 \qquad |\sigma^*(\omega_2)| = \tfrac{1}{2}\sigma_2$$

$$G^*(\omega_1) = \frac{\sigma^*(\omega_1)}{\gamma^*(\omega_1)} \qquad G^*(\omega_2) = \frac{\sigma^*(\omega_2)}{\gamma^*(\omega_2)}$$

Thus, viscoelastic information at two frequencies can be obtained by use of a single waveform.

This technique can be expanded to include n sine waves at n different frequencies, using frequency-analysis techniques to calculate $G^*(\omega_i)$. Note that $\omega_i = k \cdot \omega_0$, where k is an integer, i.e. all the higher frequencies must be integer multiples of the lowest, fundamental frequency to ensure that all frequencies are harmonically related. Any periodic waveform can, in principle, be used as a strain history.

There are, however, some limitations on the choice of the strain waveform. The amplitude of the composite strain (the sum of all the separate amplitudes) must not exceed the linear viscoelastic limit of the material being tested. At the same time, each stress component must have an amplitude large enough to be sensed by the stress or force transducer used in the rheometer. The strain amplitudes should be chosen so that the resulting stress amplitudes are all about the same magnitude. Thus, high-frequency components should usually have smaller amplitudes than those at low frequency, since they cause larger stress responses. Also, they are repeated many times during the course of the measurement, allowing the DFT to average the signal over several cycles and thus improve the signal-to-noise ratio.

Notice that in the analysis it is the *actual* measured strain that must be used rather than the commanded strain signal sent to the rheometer drive controller. These two signals are often not the same, particularly for strain waveforms with high frequency components. For example, the motor can act as a filter, reducing or even eliminating high-frequency components that are beyond its response range. The drive response at high frequencies is one of the limits of any testing system. The high-frequency response of the transducer used to measure the stress is another.

7.2.1. Time-Domain Mechanical Spectroscopy

While the main thrust of this chapter is the use of the DFT to analyze the results of DMA (dynamic mechanical analysis) experiments, another, related method for low frequencies has been developed that deserves mention. An important problem that arises in making low-frequency measurements is that the time required to perform a test is very long. At $f = 0{\cdot}01$ Hz, at least 100 s are required for a single cycle, while a frequency of $0{\cdot}001$ Hz requires at least 1000 s. Marin *et al.*[3] have developed a technique for using time-domain data, stress relaxation after cessation of steady shear (see Fig. 7.3) to obtain $G^*(\omega)$ at low frequencies. Rather than use a DFT, a Carson–Laplace transform is used to convert the time-based data to the frequency domain:

$$\eta'(\omega) = \eta_0 - \int_{T_r}^{\infty} \sigma^-(t)\cos(\omega t)\,\mathrm{d}t$$

$$\eta''(\omega) = \int_{t_r}^{\infty} \sigma^-(t)\sin(\omega t)\,\mathrm{d}t$$

(7.10)

where $\sigma^-(t)$ is the stress relaxation function, η_0 is the zero shear viscosity and t_r is the time of cessation of the shear. Numerical integration proceeds until $\sigma^-(t)$ is very small, so relaxation data must be sampled for at least this long. η' and η'' are the real and negative imaginary components of

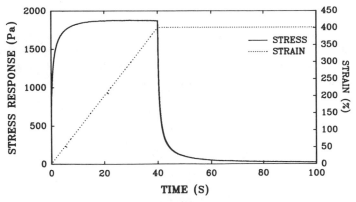

Fig. 7.3. Stress relaxation after steady-shear experiment.

the complex viscosity, η^*, and to get the complex modulus we can use:

$$G^* = j\omega\,\eta^* = j\omega\,\sqrt{(\eta')^2 + (\eta'')^2} \qquad (7.11)$$

Good agreement for η' and η'' values was found for 2–3 decades in the low-frequency region ($0.01 < \omega < 1$ rad/s, or, approximately $0.0016 < f < 0.16$ Hz) for several thermoplastics.

7.3. PROPERTIES OF THE DISCRETE FOURIER TRANSFORM

The dynamic response of a system can be described in the time domain as a function of time or in the frequency domain as a function of frequency. These are two methods of representing the same information, and the Fourier transform is used to convert functions from one representation to the other.

The discrete Fourier Transform (DFT) is related to the Fourier transform, but it is defined only for discrete values over a finite interval. The DFT provides the **exact** transform for the discrete values provided. The results are typically interpreted from a continuous viewpoint, however, and while this is usually quite straightforward, one must be aware of the differences between continuous and discrete perspectives so as to minimize errors in interpretation.

The DFT converts a discrete, complex-valued, time domain set of data to a discrete, complex-valued, frequency-domain series, and is defined as follows:

$$X(k) = \frac{1}{N} \sum_{n=0}^{N-1} x(n)\,e^{-j2\pi kn/N} \qquad (7.12)$$

where x is the time based signal, X is the transformed signal, N is the number of data points, $j \equiv \sqrt{-1}$, and k is the kth element in X. Using Euler's identity, X can also be expressed as

$$X(k) = \frac{1}{N} \sum_{n=0}^{N-1} x(n)\left[\cos\left(\frac{2\pi kn}{N}\right) - j\sin\left(\frac{2\pi kn}{N}\right)\right] \qquad (7.13)$$

Fourier analysis is based on the Fourier series representation of a function,

$$x(t) = a_0 + \sum_{i=1}^{\infty} [a_i \cos(i\omega_0 t) + b_i \sin(i\omega_0 t)] \qquad (7.14)$$

where $\omega_0 = 2\pi/T$, $T = $ period of the waveform $= 1/F$, and $F = $ the fundamental frequency. Note that an infinite series is used to represent continuous functions, while a finite series is used for discrete functions (such as sampled data). For example, a continuous square wave has the following infinite series representation:

$$x(t) = \frac{4V}{\pi} [\cos(\omega_0 t) - \tfrac{1}{3} \cos(3\omega_0 t) + \tfrac{1}{5} \cos(5\omega_0 t) - \tfrac{1}{7} \cos(7\omega_0 t) + \cdots]$$

Fourier analysis assumes that the waveform being studied is periodic, i.e. that $x(t) = x(t + \tau)$, where τ is the period of the waveform, and that it extends in time from negative to positive infinity.

Of course, this assumption is not valid for physical systems, since these must have specific starting and ending times. Startup transients in the input strain and the stress response will cause aperiodicity in the measured signals at the start of a test. In Fig. 7.4, three cycles of a complex strain history are shown, along with the resulting stress response. The first three peaks of the initial cycle differ considerably from the corresponding peaks in the following cycles due to startup transients. This transient response seems to die out about half way through the first cycle, so that by the second cycle a pseudo-steady state is achieved. Repeating the waveform until any transients have become negligible before sampling data is sufficient to ensure that the assumption of periodicity is valid. If the waveform has a very long period, only part of it may need be repeated,

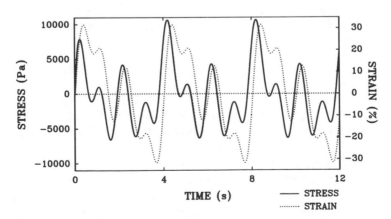

Fig. 7.4. Start-up transients in a dynamic measurement using a multi-sinusoid composite waveform.

depending on how quickly the transients die out. As a rule of thumb, the strain should be applied for a length of time greater than the longest relaxation time of the material being tested before recording any data points. Stress relaxation experiments (monitoring the stress response to a step in strain) can be used to get an approximate idea of the longest relaxation time involved.

Another important property of the DFT is that it is linear. A component added in the time domain is added in the frequency domain. This allows the construction of a waveform with a desired frequency spectrum by adding together appropriate waveforms in the time domain. Each waveform adds its own components until the desired frequency spectrum is achieved.

The calculation of the DFT is easily accomplished using a computer by simply taking the sum of the terms shown in eqns 7.12 or 7.13. However, this can be very time consuming, even on a fast computer. A number of algorithms called fast Fourier transforms (FFT) have been developed that greatly speed the calculation.[4] To calculate the DFT, the number of mathematical operations required is proportional to N^2 whereas the FFT algorithm requires $N \cdot \log_2(N)$ operations. This is much less, especially when N is large. A limitation of the FFT is that it generally requires the data series to be a power of two in length (i.e. to have 2, 4, 8, 16, 32, . . . elements).

If it is desired to increase the signal-to-noise ratio by sampling a large number of cycles, it is usually better to average the cycles first, and then transform, rather than allowing the transform to do the averaging. Even using FFT algorithms, the time necessary for calculations increases rapidly as the number of points increases, and by averaging first one reduces the number of data points.

7.3.1. Aliasing

In comparing discrete and continuous data, a factor of central importance is the sampling rate. If this rate is higher than necessary, surplus data points will be collected, resulting in long calculation times and excessive data-storage requirements. If the sampling rate is too low, not enough information is collected and *aliasing* occurs.

The Nyquist sampling theory (known also as Shannon's theory) indicates that the sampling frequency, f_s, must be at least twice that of the highest frequency component in the signal being sampled. Conversely, the Nyquist frequency, f_{Nyq}, is defined as the highest frequency sinusoid that can be resolved at a given sampling rate. If Δt is the sampling period,

$$f_{\mathrm{Nyq}} = \tfrac{1}{2} f_\mathrm{s} = \frac{1}{2\Delta t} \tag{7.15}$$

Aliasing is the association of a high-frequency component with an incorrect, lower frequency and occurs when a signal contains components having frequencies above f_{Nyq}. These higher harmonics are folded back about f_{Nyq} into the frequency domain below, adding to the valid components already there. For example, if one were to sample a 100 Hz sine wave at $f_\mathrm{s} = 120$ Hz, aliasing would occur. Since $f_{\mathrm{Nyq}} = 60$ Hz, the 100 Hz component would appear at 20 Hz, reflected back from f_{Nyq}. The 100 Hz component is 40 Hz higher than f_{Nyq} and is reflected backwards from 60 Hz by 40 Hz to the 20 Hz component.

Figure 7.5 shows what happens in the time domain when a 100 Hz sinusoid is sampled at a frequency of 95 Hz. The result is a spurious 5 Hz signal. One can usually tell if aliasing is occurring by inspecting the frequency spectrum generated by the DFT. If the magnitude of the harmonics has not dropped to zero (or very close) before reaching the Nyquist Frequency, then aliasing is probably occurring, and a higher sampling rate must be employed to represent the sampled signal accurately

When using an arbitrary waveform to generate a strain having components in a certain range of frequencies, care must be taken when setting the sampling rate if there are significant harmonics outside the desired bandwidth. The sampling frequency must be set high enough to resolve

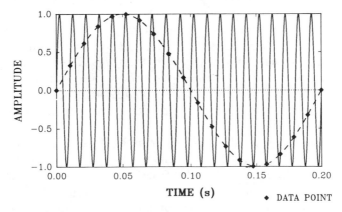

Fig. 7.5. Spurious 5 Hz sine wave from sampling a 100 Hz signal at $f_\mathrm{s} = 95$ Hz.

accurately the undesired harmonics as well or they will be aliased back into the desired range. For example, the DFT of the PRBS function (see Section 7.4.4) has a $|\sin(x)/x|$ shape. Usually only the first few humps of the transform are of interest, but the higher-frequency harmonics must also be resolved when sampling or they will reflect back from f_{Nyq} and pollute the information in the desired range.

Corresponding to the DFT there is an inverse function, IDFT, that transforms data from the frequency domain to the time domain. It is very similar to the forward transformation

$$x(n) = \sum_{k=0}^{N-1} x^*(k)\, e^{j2\pi kn/N} \tag{7.16}$$

i.e. the transform of the complex conjugate of $x^*(\omega)$ without the $1/N$ scaling constant. The IDFT will return the time-base data from the transformed data to within round-off error, provided that there has been no aliasing. If aliasing has occurred, high-frequency components will be missing from the restored time function, which will also have low-frequency amplitudes greater than it should. The information above the Nyquist Frequency is thus lost permanently, and the original signal cannot be recovered.

7.3.2. Time- and Frequency-Domain Scaling

Scaling in the time and frequency domains are closely related. Fixing the sampling period, Δt (or its inverse, $f_s = 1/\Delta t$), and the number of samples collected, N, sets the ranges of both the time and frequency axes. The time axis runs from 0 to T, where $T = (N-1)\Delta t$, in increments of Δt. The frequency axis runs from 0 to f_{Nyq}, where $f_{\mathrm{Nyq}} = \frac{1}{2} f_s$, in increments of Δf, where $\Delta f = f_s/N$. Increasing the resolution in one domain causes a decrease in the other, if N is held constant; as Δt is decreased to get better time-domain resolution, Δf will increase. Two units of frequency are in common use; cycles per second, f (Hertz), or radial frequency, ω (rad/s). Each has advantages in certain situations, and they can be employed interchangeably with $\omega = 2\pi f$.

Regarding the frequency axis, different DFT (or FFT) subroutines return the transform in different formats. When given an N-point time series, most will return an N-point frequency series with the frequency axis in one of two formats: either ranging from $-f_{\mathrm{Nyq}}$ to $+f_{\mathrm{Nyq}}$, or from 0 to $(N-1)\Delta f$. These two formats contain the same information, except that the negative-frequency axis information in one has been mapped to frequencies above the Nyquist frequency in the other. This is because it is easier to map N points on to 0 to $N-1$ than $-N/2$ to $N/2$.

For any real-valued time signal, the magnitude of the transform in the negative-frequency domain is a mirror image of that in the positive domain, and the phase is duplicated but inverted. We are always dealing with real-time signals, so only the range 0 to f_{Nyq} is of interest; the other half of the transform contains a duplication of information. Some DFT routines assume a real input signal and only return the first $N/2+1$ points, corresponding to the range 0 to f_{Nyq}, eliminating the duplication.

This duplication of information is why the amplitude of a frequency harmonic is only half the value of the amplitude in the time domain. Half of the amplitude appears in the positive-frequency domain, and half appears in the negative domain.

7.3.3. Leakage

When setting the data-acquisition parameters for an experiment (sampling rate and number of points), care must be taken to ensure that an integral number of cycles will be sampled. The length of the cycle is then harmonically related to the length of the sampling window, and the frequency components of the signal fall exactly on the discrete points in the DFT. When a non-integral number of cycles is sampled, the actual frequency of a component falls between two of the discrete frequency points, and the amplitude is split between them. As a result, the amplitude is reduced and spread over more than one frequency node. This phenomenon is called leakage. Leakage can also have a considerable effect on the phase of the frequency signal.

A spurious discontinuity can be introduced into a waveform if a non-integral number of cycles are sampled, as a result of the assumption of the periodicity of the signal. The last points in Fig. 7.6(a) and (b) (hollow symbols) are the first points in the next cycle if periodicity is assumed. Since the sampled waveforms (a) and (b) both start at zero, the first point of the next waveform must also start at zero. In Fig. 7.6(b), we see that this introduces a discontinuity. If an integral number of cycles is sampled, then the end of one cycle coincides with the start of the next, and the waveform is continuous. The DFT assumes that the window of data sampled is periodic, so it is possible to get apparent discontinuities even with periodic signals. If one is forced to sample non-periodic signals, special 'windowing' techniques are available that can reduce the problems encountered due to leakage.[5]

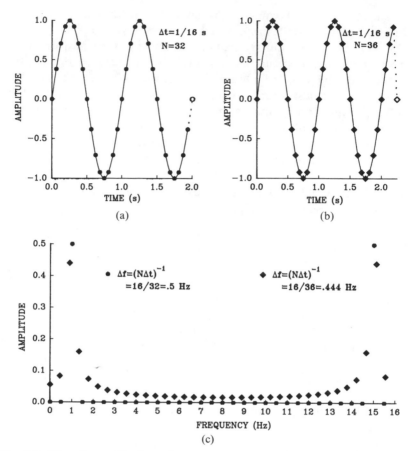

Fig. 7.6. The effects of leakage due to assumed periodicity. a) Integral number of cycles; b) Non-integral number of cycles; c) DFT's of time based data.

Figure 7.6(c) shows a comparison of the transforms of the waveforms shown in Fig. 7.6(a) and (b). The circles (integral number of cycles) have only a single harmonic at 1 Hz (mirrored at 15 Hz, since $f_{Nyq} = 8$ Hz) with all other harmonics equal to zero. The diamonds show a reduced peak at $f = 0.888$ Hz, and small but appreciable harmonics across the entire spectrum. Thus, the sudden discontinuity introduces extra-high-frequency components, while the transform's inability to represent 1 Hz ($\Delta f = 0.444$ Hz) exactly causes that peak to be spread out over neighbouring nodes.

When using non-periodic waveforms (i.e. step strains, or the equi-strain waveform of Section 7.4.2), the actual strain motion must duplicate any spurious discontinuities due to assumed periodicity. If it does not, the discontinuity will appear in the transform of the strain signal with no matching harmonics in the stress signal, leading to incorrect results. The strain signal must include a steep ramp to represent the discontinuity, so that the necessary stress spike is generated. See Section 7.4.2 for an example of this phenomenon.

7.3.4. Alternating *versus* Simultaneous Data Acquisition

Care must be taken during data acquisition to ensure that both the stress and strain signals are sampled at the same time. The frequency-analysis techniques assume that this is true; if it is not, an erroneous phase shift between the stress and strain will be introduced, which will affect the value of δ obtained.

Most data-acquisition boards that do not have simultaneous sample and hold for all channels, perform sampling in an alternating fashion. The board runs at twice the desired sampling frequency (for two channels) and scans the channels in succession, retrieving data at the desired rate but not at the same instant. The phase shift due to this method of sampling can easily be calculated and a correction applied to the data.

Consider two waveforms at the same frequency and separated by a phase shift δ, sampled alternately, as shown in Fig. 7.7. The 'x' wave is

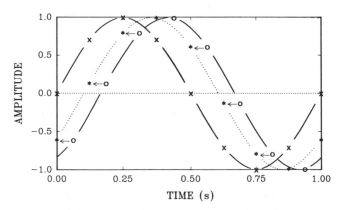

Fig. 7.7. Relative phase shift due to alternating sampling. The 'o' wave is shifted to the position of the '*' wave, reducing the phase difference between the 'x' and 'o' waves.

$\sin(2\pi f_0 t)$, while the 'o' wave is $\sin(2\pi f_0 t - \delta)$, with $f_0 = 1$ Hz. The data acquisition system is sampling 'xoxoxoxoxo...' at $f_a = 2 \cdot f_s$, twice the desired sampling frequency. Since the DFT calculation assumes simultaneous sampling, the 'o' wave is shifted to the left by one alternating sample period to line up with the 'x' wave (to the position of the dotted-line '*' wave). One alternating period is the equivalent of one half of a simultaneous period, so that the additional phase shift δ_a in rad/s is

$$\delta_a = 2\pi \frac{\frac{1}{2}\Delta t}{T_0} = \pi \Delta t f_0 = \pi \frac{f_0}{f_s} \tag{7.17}$$

where $\Delta t = 1/f_s$ is the simultaneous sampling period, $T_0 = 1/f_0$ is the period of the sine wave, and $\sin(2\pi f_0 t - \delta + \delta_a)$ is the equation of the '*' wave. The phase shift δ_a depends upon the sampling rate, f_s, and the frequency of the waveform being sampled, f_0.

The simplest way to apply the correction given by eqn (7.17) is to transform whichever waveform is being sampled second (the 'o' wave in Fig. 7.7) to the frequency domain in polar form and subtract the correction from the phase of the waveform at each frequency. While f_s is held constant at the value used during data acquisition, the f_0 value used is the frequency of the point being corrected; δ_a is a linear function of frequency, so that its value is increasing as it is subtracted from the higher-frequency phase components. The waveform can then be converted back to Cartesian coordinates or restored to the time domain. Only one of the signals should be so treated, since it is the relative phase between the signals that is being corrected. Note also that the signal frequency and the sampling frequency must be in the same units, either Hz or rad/s; mixing units will bring an extra factor of 2π to the top or bottom of the equation.

7.4. SOME WAVEFORMS OF SPECIAL INTEREST

While in theory any periodic waveform can be used to generate a strain history for dynamic measurements, some waveforms are more useful than others. Good waveforms have either a flat frequency spectrum with a high amplitude over as wide a range as possible, or a number of sharp peaks at specific frequencies of interest. We describe below some of the functions that have been proposed for the measurement of linear viscoelastic material properties.

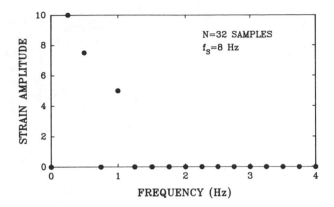

Fig. 7.8. Transform of three component multiple sine wave signal.

7.4.1. Multiple Sine Waves

As was discussed in Section 7.2, a complex waveform can be built up by adding together two or more sine waves of different frequency that are all integral multiples of a base frequency. The frequency content of this waveform is limited to the discrete frequencies of the component sine waves. The cyclic signal in Fig. 7.4 is composed of three sine waves, with a fundamental radial frequency $\omega_0 = 2\pi$ rad/s ($f = 1$ Hz) and is described by the following equation:

$$\gamma(t) = 20 \sin(\omega_0 t) + 15 \sin(2\omega_0 t) + 10 \sin(4\omega_0 t)$$

where $\gamma(t)$ is the strain in percent. Figure 7.8 is the frequency-domain representation of the function, which has non-zero harmonics corresponding to the three component frequencies; this waveform can be used to determine G^* at these three frequencies.

Holly *et al.*[6] used this method to measure the dynamic moduli of both non-reacting and reacting polymers at several frequencies simultaneously. Also, several commercial rheometers that measure dynamic properties now offer this capability as an option.

7.4.2. Equi-strain Waveforms

These waveforms, suggested by Malkin *et al.*,[7] have a uniform strain amplitude across a specified band of frequencies. The signal has the form

$$\gamma(t) = \gamma_0 \frac{\sin(\omega_0 t)}{\omega_0 t} \tag{7.18}$$

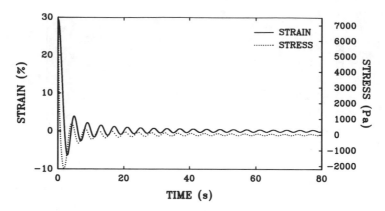

Fig. 7.9. Equi-strain waveform, $\omega_0 = \pi/2$ rad/s, $\gamma_0 = 30\%$.

which is displayed in Fig. 7.9 for a cut-off frequency of $f = 0.25$ Hz, or $\omega_0 = \pi/2$ rad/s. Equation (7.18) gives a constant amplitude for each frequency in the spectrum less than ω_0, as shown in Fig. 7.10. The flat region below 0.25 Hz is over a decade in size.

Care must be taken when using this waveform since it is not periodic. Equation (7.18) has a value of $\gamma = \gamma_0$ at $t = 0$, with γ approaching zero as t becomes large. This discontinuity (between the starting and finishing values of the waveform) introduces high-frequency harmonics and the possibility of aliasing, since the harmonic amplitudes in Fig. 7.10 are still falling off rapidly at $f = f_{\text{Nyq}}$ (in this case, 5 Hz). Windowing techniques

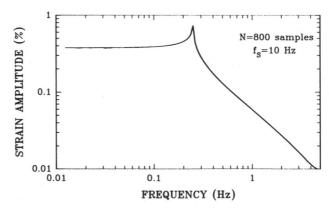

Fig. 7.10. Transform of the equi-strain function, $\omega_0 = \pi/2$ rad/s.

can be used to reduce these effects if they become significant.[5] The discontinuity present in the strain must also be reflected in the stress measurement. The strain history must have a step from 0 to 30% at time $t = 0$ in order that the stress measurement has a spike to match the effects of the discontinuity in the strain. In other words, since the DFT assumes the strain discontinuity is there, it must be there in actual fact in order that the stress reflects its presence. In Fig. 7.9, the stress has a large positive spike at time $t = 0$ that would not be present without the initial step in the strain; without this spike, incorrect results would be obtained.

7.4.3. Exponential Shear

A waveform suggested by Bertig *et al.*[8] consists of an exponential strain of the form

$$\gamma(t) = \gamma_0 \left(\frac{\alpha}{\pi}\right)^{\frac{1}{2}} \exp[-\alpha(t - t_{\text{offset}})^2] \qquad (7.19)$$

As can be seen from Fig. 7.11, this strain history is very close to a pulse (in the combined function, two pulses), so that its transform will have a flat amplitude over a wide range of frequencies. γ_0 is used to scale the size of the pulse, while α is a 'sharpness' parameter that determines the speed of the rise and fall of the strain. Also, α has an effect on the range of frequencies contained in the waveform, since the sharper the rise and fall, the larger the number of high-frequency harmonics that will be present. The time at which the pulse occurs is governed by t_{offset}. In Fig. 7.11,

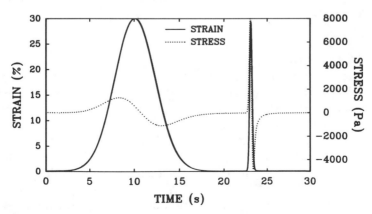

Fig. 7.11. Combined exponential shear functions, with $\alpha = 0{\cdot}1$ and $t_{\text{offset}} = 10$ s for the broad pulse, $\alpha = 25$ and $t_{\text{offset}} = 23$ s for the sharp pulse.

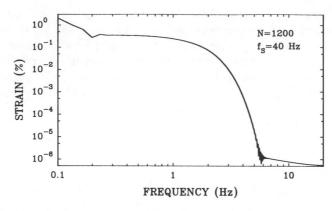

Fig. 7.12. Transform of combined exponential strain function, $\alpha = 0 \cdot 1$ and $t_{offset} =$ 10 s for the low-frequency pulse, $\alpha = 25$, $t_{offset} = 23$ s for the high-frequency pulse.

two pulses with different α values (i.e. different frequency contents) are strung together into a single test to obtain a broader bandwidth. The first pulse has $\alpha = 0 \cdot 1$ and $t_{offset} = 10$ s, while the second, higher-frequency pulse has $\alpha = 25$ and $t_{offset} = 23$ s.

The transform of the exponential function also has an exponential shape. There is a sharp high-frequency cut-off, reducing high-frequency components beyond the range of interest. This means fewer worries about high-frequency aliasing; lower sampling frequencies and fewer data points can be used. The transform in Fig. 7.12 has two humps, one at very low frequencies and a higher one, corresponding to the two pulses in Fig. 7.11.

7.4.4. PRBS Waveforms

A pseudo-random binary sequence (PRBS, sometimes called a binary maximum length sequence, or BMLS) is a waveform commonly used in stochastic system identification to obtain an estimate of a system's impulse response. It has almost the same autocorrelation as white noise (a single impulse at lag 0) but is cyclic, so that only one period of the waveform need be sampled to obtain all necessary information. It has also a fairly flat frequency spectrum, the range of which can be controlled through the parameters of the PRBS, n_B and Δt_B.

The PRBS is a series of steps up and down between two signal levels. It is periodic with a period T_B, and can change levels only at certain discrete times, $k\Delta t_B$, where k is an integer and Δt_B is the minimum discrete time interval for a step. The PRBS consists of a chain of 'up' and 'down' states of different lengths, with the number of 'up' states roughly equal to the number of 'down' states. The total number of discrete time intervals, N_B, is a function of the 'degree' n_B of the PRBS, where

$$N_B = 2^{n_B} - 1$$
$$T_B = N_B \cdot \Delta t_B$$

(7.20)

Since N_B is always odd, there is always one more 'up' state than 'down'. See Davies[9] for details on the generation of a PRBS, and Fig. 7.13 for some sample waveforms.

The frequency spectrum of a PRBS has a $|\sin(x)/x|$ envelope, as shown in Fig. 7.14. The 3 db drop-off occurs at approximately $f_{max} = 1/(3\Delta t_B)$ Hz, giving the PRBS a useful range conservatively estimated from $f_{min} = 1/(N_B \cdot \Delta t_B)$ to f_{max} (in many cases, the PRBS gives useful results far beyond the suggested f_{max} limit, as in the sample experiment in Section 7.5). The amplitude reaches zero for the first time at $f = 1/\Delta t_B$.

The drive systems that provide the strain for dynamic mechanical measurements cannot provide the infinite shear rate necessary to generate the steps required for an ideal PRBS. Most systems can, however, provide a steep ramp. In fact, such high-speed ramps are used to determine the relaxation spectrum in the time domain. Thus, the PRBS can be thought

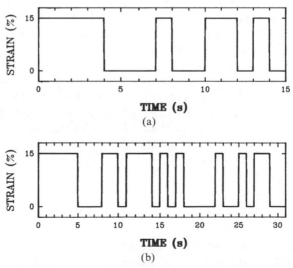

Fig. 7.13. Sample PRBS waveforms for $n_B = 4$ and 5.

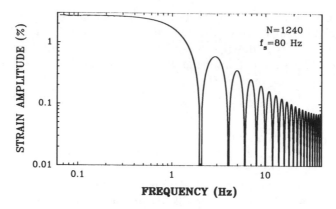

Fig. 7.14. DFT of a PRBS, $n_B = 5$, $\Delta t_B = 0.5$ s.

of as a series of stress-relaxation experiments, repeated at varying intervals to obtain the desired frequency spectrum; Vratsanos and Farris[10] used step strains and Fourier transforms to generate viscoelastic data. The drive motor will likely cause some irregularities in the frequency spectrum. However, as long as the *actual* strain (as opposed to the commanded strain) is used in the analysis of the data, there should be little effect on the final result. If the high-frequency harmonics are not desired and are causing aliasing problems, filtering the waveform can reduce their amplitudes. The filtering may be performed either digitally when programming the waveform generator or directly on the analog control signal as it is sent to the rheometer drive controller.

A signal that resembles the PRBS is a randomly switched waveform. After each time point $k \cdot \Delta t_s$ (where k is an integer), the signal has a certain probability of switching to the opposite state. This is shown in Fig. 7.15, where the probability of switching at a given point is 50%. This waveform is simpler to generate than a PRBS but does not have the same well-defined statistical properties, particularly if a small number of steps is used. In Fig. 7.15, the mean 'up' time is 38·7%, considerably lower than the ideal value of 50%.

7.5. A SAMPLE DMA EXPERIMENT

We now describe in detail the design and analysis of an actual experiment using a PRBS strain signal. The polymer sample was a silicone putty (polydimethylsiloxane) having room-temperature properties that are

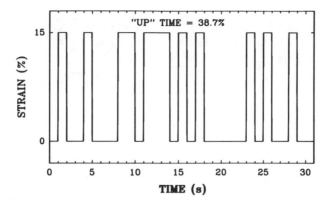

Fig. 7.15. Randomly switched waveform, $p_s = 50\%$, $N_s = 31$, $\gamma_s = 15\%$.

similar to those of many thermoplastic melts. A rotational rheometer (Rheometrics, Piscataway, New Jersey, USA, RDA II with the arbitrary waveform option and an external waveform generator) with 25-mm diameter parallel disk fixtures was used. The gap was 1·5 mm, and the temperature was 30°C.

First we select the frequency range over which results are desired and determine the PRBS parameters, n_B and Δt_B that will provide harmonics in this range. In general, we choose a value of Δt_B first to fix the upper end of the frequency range. The width of the range is then governed by the length of the PRBS, i.e. by n_B, which determines how many steps will be taken. In this case, a bandwidth of 0·06–0·6 Hz, corresponding roughly to 0·4–4 rad/s, was chosen. The PRBS will expand the high-frequency end of this range if we use the conservative frequency max/min estimates of Section 7.4.4 to calculate parameters. Once the frequency range has been selected, theoretical values for the minimum discrete time step Δt_B and degree n_B are determined using eqn (7.20). Convenient values are then chosen to match theoretical values as closely as possible:

$$f_{\max} = \frac{1}{3\Delta t_B} = 0\cdot 6 \text{ Hz}$$

$$\Delta t_B = 0\cdot 5555 \text{ s} \quad \Rrightarrow \quad \Delta t_B = 0\cdot 5 \text{ s}$$

$$f_{\min} = \frac{1}{N_B \Delta t_B} = 0\cdot 06 \text{ Hz}$$

$$N_B = 33\cdot 3 \quad \Rrightarrow \quad n_B = 5, \quad N_B = 31 \text{ steps}$$

In this case, Δt_B is set at 0·5 s, close to the theoretical value of 0·5556 s. The theoretical value of N_B, which is 33·3, leads to selecting $n_B = 5$, giving $N_B = 31$ steps. The frequency bandwidth calculated for the test parameters is $f_{min} \approx 0·065$ Hz and $f_{max} \approx 0·667$ Hz, close to the desired range. The size of the step strain, γ_B, was chosen to be 30%, which is in the linear viscoelastic region for this polymer, and gives a reasonable stress pulse amplitude.

Once the strain signal is fixed, the data sampling rate must be established. There are several criteria that f_s must satisfy.

(1) The frequency selected must be high enough to resolve all of the significant high-frequency harmonics in the PRBS to avoid aliasing.

(2) To avoid leakage, each step in the PRBS should be sampled an integral number of times.

To sample the harmonics at f_{max}, f_{Nyq} must be at least 0·667 Hz. Figure 7.14 shows the transform of a sample PRBS with $n_B = 5$ and $\Delta t_B = 0·5$ s; it is obvious that a much higher sampling rate than $f_s = 2f_{max} = 1·333$ Hz will be required. For this particular test, $f_s = 80$ Hz was chosen, giving $f_{Nyq} = 40$ Hz. While the harmonics in Fig. 7.14 have not quite died out by the time they reach 40 Hz, they are small enough to have little effect. Also, the region polluted by any aliasing will be in the range 20–40 Hz, well above the range of interest. In addition, the high-frequency harmonics will tend to be damped by the motor drive in the actual strain signal, further reducing any aliasing. The control computer is capable of storing up to 1600 points per experiment; with $\Delta t_B = 0·5$ s, 40 points per step will be sampled, 1240 points for the entire waveform.

When applying the PRBS strain to the sample, care must be taken to ensure that all start-up transients have died out before starting to sample data. The results of a time-domain stress-relaxation experiment will help us to get some idea of the time required for the fluid to adapt to a sudden change in strain. The stress-relaxation curve for the silicone putty at T = 30°C is displayed in Fig. 7.16. The stress drops off fairly rapidly, and is very close to zero after 6 s. Allowing a few additional seconds as a margin for error, we wait for 15 s after starting the strain waveform before starting data acquisition. Timing the start of acquisition to coincide with the start of the PRBS is not necessary. As long as an integral number of PRBS cycles is sampled, the beginning and end of the sampled waveform will be continuous.

Once data for the experiment have been collected, they must be converted to proper units (in this case, stress in Pa and strain in strain units

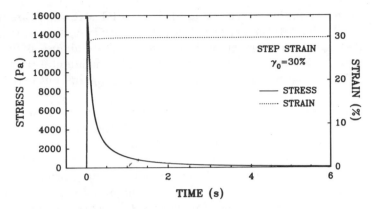

Fig. 7.16. Stress-relaxation experiment; silicone putty at 30°C.

(%)). Figure 7.17 shows the time-domain results. There is a slight strain distortion in each shear step; the initial part of a step is sharp, but the top is slightly rounded by the motor response. This will slightly reduce the high frequency content of the signal. The stress response looks like a series of stress relaxations similar to Fig. 7.16.

If more than one cycle had been sampled, the data would be averaged to reduce the signal noise and the number of points for the DFT. A DFT subroutine is then used to transform $\sigma(t)$ and $\gamma(t)$ into the frequency domain. Since the number of samples is not an even power of two, a

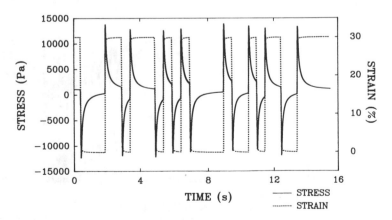

Fig. 7.17. PRBS test: stress and strain results (time domain).

normal FFT algorithm cannot be used. Some FFT algorithms are now available (MATLAB, Natick, MA, USA, V3.5's FFT function, was used here) that accept any number of data points and are almost as fast as regular FFT subroutines. In all cases, a straight summation of the terms of eqn (7.13) could be used; it is very simple to program, but may take a while to run.

For this experiment, alternating data sampling was used, with $\sigma(t)$ sampled first and $\gamma(t)$ second. The correction mentioned in Section 7.4.2 should be applied to the strain. After transforming $\gamma^*(f)$ to polar coordinates ($|\gamma^*|$ and phase in radians), we subtract $\delta_a = \pi f/f_s$ from each phase value, where f is the frequency of the component being corrected and f_s is the sampling frequency, 80 Hz. This is an important correction, as δ_a can become significant at higher frequencies. Corrections for the torque transducer response, if any, (i.e. compensation for phase lags at higher frequencies) should also be applied here as well.

Examination of the magnitudes of $\sigma^*(f)$ and $\gamma^*(f)$ in Fig. 7.18 show that the higher-frequency harmonics fall to $1/200$ of their low-frequency amplitudes before reaching $f_{Nyq} = 40$ Hz. This suggests that the results should be free of any aliasing effects. Notice that any fluctuation in the amplitude of $\gamma^*(f)$ is matched by a similar blip in $\sigma^*(f)$, so that the ratio between them, $G^*(f)$, is a smooth function.

Point-by-point complex division of the $\sigma^*(f)$ function by $\gamma^*(f)$ gives the function $G^*(f)$, displayed in Fig. 7.19 in polar form. The magnitude and phase of G^* are valid beyond the original estimate of f_{max}. The PRBS data compare well with the data obtained using single-frequency tests

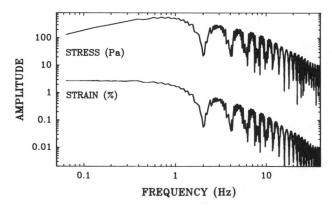

Fig. 7.18. Stress and strain: frequency domain results.

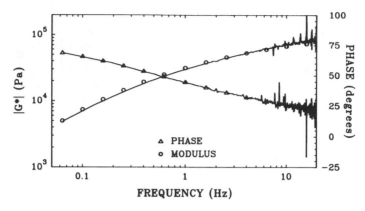

Fig. 7.19. PRBS experiment results compared with single-point data.

with the same sample, with acceptable data obtained over two decades. At the high-frequency limit, the signals become very noisy; here the strain amplitudes have become so small that the stress and strain measurements become lost in system noise. Note that the time necessary to perform the experiment is much less than that required to generate single-point measurements over the same frequency range.

This test was performed with $f_s = 80$ Hz. If a lower sampling frequency, say $f_s = 40$ Hz, had been used, aliasing problems would have been encountered. The results of an experiment identical to the one described above, except with $f_s = 40$ Hz, are shown in Fig. 7.20. Although the amplitude of G^* appears to be satisfactory, the phase drifts upward beyond 1 Hz; this is due to aliasing. Near $f_{Nyq} = 20$ Hz, the signal amplitudes are very low compared to their low-frequency counterparts; however, they are still comparable to one another and cause aliasing in the higher frequencies. Phase, as well as amplitude, can be affected by aliasing, and in this case the phase information between 1 Hz and $f_{Nyq} = 20$ Hz has been contaminated. Waveforms with sharp high-frequency cut-offs, such as the exponential waveforms presented in Section 7.4.3, or a filtered PRBS strain with high-frequency harmonics removed, can ease this problem by eliminating the high-frequency components and limiting any aliasing.

If a larger-frequency bandwidth is desired for an experiment (i.e. decreasing f_{min} to get more low-frequency information), the length of the PRBS, N_B (or n_B), must be increased. Since we have shown that f_s cannot be decreased without aliasing problems, the number of data collected must be increased (i.e. keep the same sampling rate, but sample over a

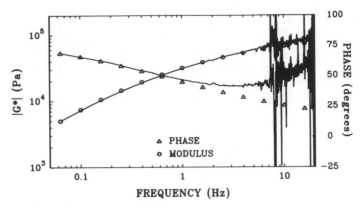

Fig. 7.20. PRBS test results for $f_s = 40$ Hz experiment; effects of aliasing.

longer period of time). This can lead to problems on instruments without sufficient data-storage capability; thus the number of samples a machine is capable of storing is a major variable in determining the frequency bandwidth of a single experiment.

REFERENCES

1. G. R. Zeicher and P. D. Patel, *Proc. 2nd World Congress of Chemical Engineering, Montreal*, 1981, **6**, 373.
2. W. H. Tuminello, *Poly. Eng. Sci.* 1986, **26**, 1339.
3. G. Marin, J. Peyrelasse and Ph. Monge, *Rheol. Acta*, 1983, **22**, 476.
4. J. W. Cooley and J. W. Tukey, *Math. Computing*, 1965, **19**, 297.
5. R. W. Ramirez, *The FFT—Fundamentals and Concepts*, Prentice Hall, Englewood Cliffs, NJ, 1985.
6. E. K. Holly, S. K. Venkataraman, F. Chambon and H. H. Winter, *J. Non-Newtonian Fluid Mech.*, 1988, **27**(1), 17.
7. A. Ya. Malkin, V. P. Beghishev and V. A. Mansurov, *Vysokomol. Soedin* (J. Polymer Science USSR), 1984, **26A**(4), 869.
8. J. P. Bertig, J. J. O'Connor and M. Grehlinger, Use of arbitrary waveforms to determine the rheological properties of viscoelastic systems. Paper presented at Society of Rheology 61st Annual Meeting, 1989.
9. W. D. T. Davies, *System Identification for Self-Adaptive Control*, Wiley Inter-Science, New York, 1970.
10. M. S. Vratsanos and R. J. Farris, *J. Appl. Polym. Sci.*, 1988, **36**, 403.

Chapter 8

Rheological Measurements on Small Samples

MICHAEL E. MACKAY

*Department of Chemical Engineering, The University of Queensland,
St Lucia, Brisbane, Queensland, Australia*

8.1. INTRODUCTION

'Rheological Measurements on Small Samples' is a somewhat ambiguous title. The question arises, How small is small? We take it that the use of rheological data measured on small samples will be for designing or interpreting bulk-flow processes where the fluid may be considered a continuum. Thus, small means the smallest sample one can use that accurately represents the bulk rheological properties. In some cases the effect of sample size on the rheological properties is needed and will be discussed below. Pragmatically, a small sample is considered to have a volume in the range 1–20 µl, since sample loading becomes difficult for volumes below this range. The maximum volume of 20 µl is chosen, as it is the volume of a typical polymer pellet. A 1 µl volume is equivalent to a cube with 1-mm long sides and 20 µl, a cube with 2·7 mm sides. The rheological properties in shear flow of small samples is exclusively discussed in this chapter.

One of the few papers devoted to the measurement of the rheological properties of small samples is by Benbow and Lamb[1] who measured the rheological properties of polymer melts with miniature capillary and cone-and-plate rheometers. Also, Dintenfass[2] has listed several miniature rheo-meters used for biological materials.

There are three main reasons for measuring the rheological properties of small samples. The first is that only a small amount of sample may be available. This is a problem often encountered for biological systems as frequently only a small amount of sample can be synthesised or gathered. This can also be a problem when experimental polymers are synthesised and only one gram of material is available. All of this sample would be required for rheological characterisation with a standard rheometer prohibiting other tests from being performed.

The second reason for measuring the rheological properties of small samples is to determine the influence of the confining walls containing the fluid. This has applications in flow through small pores and capillaries, drainage of thin films and in lubrication.[3] The study of wall effects is achieved by using the surface-forces apparatus of Israelachvili and Tabor,[4] which has proved a very useful apparatus for measuring rheolog-ical properties of fluids at gaps of the order of nanometers. It has been shown that the rheological properties change considerably when very small gaps are used and this will be discussed below. These first two reasons are intimately linked, since, according to the first, one wishes to determine the smallest possible sample one can use to measure true bulk properties of a homogeneous fluid without wall effects.

In some cases, the variation in rheological properties with volume is needed. This constitutes the third reason. For example, local concentra-tion gradients over a finite volume may be present in a blend of compo-nents, and the rheological properties measured depend upon the amount of sample. Presumably the rheological properties will eventually become volume-independent. A blend may be two liquids (i.e. two polymer melts or an emulsion), a solid and liquid (i.e. a suspension) or a gas and liquid (i.e. a foam). The variation of rheological properties with sample size for inhomogeneous materials will not be discussed further in this chapter.

8.2. MINIATURE TORSIONAL RHEOMETERS

Several miniature rheometers have been used to measure the rheological properties of small samples. A torsional rheometer is one where the sample is deformed between two surfaces with the one surface rotating at a rate Ω relative to the other. Standard attachments used are the cone and plate, parallel plates and concentric cylinders (Fig. 8.1). Errors associated with geometry effects (i.e. gap set errors, etc.) are discussed below; however, the torque and normal force measurement will be subject to error since they will decrease in magnitude upon miniaturisation. It is assumed that this error is overcome by using much more sensitive measurement techniques.

8.2.1. Cone and Plate
The truncated cone and plate geometry shown in Fig. 8.1(a) consists of a cone that makes an angle α_{cp} with the plate. The cone rotates with respect to the plate and the torque (M) and normal force (F_n) are measured. The truncation is usually quite small $(C \sim 50\ \mu m)$ and serves the purpose of preventing tip dulling and the possibility of the pressure tending to infinity at the centre. Upon miniaturisation the truncation height will certainly be made smaller than the above value; the discussion below assumes that it is kept at 50 μm to highlight the effect of the truncation.

This geometry is attractive as the shear rate is approximately constant throughout the sample and seems a likely candidate for miniaturisation. However, the assumption that the shear rate is constant is subject to criticism, particularly when dealing with small samples, and will be addressed below.

The volume of sample contained between the cone and plate is given by,

$$V = \tfrac{2}{3}\pi R_{cp}^3 \tan{(\alpha_{cp})} \left\{ \frac{3}{2} - \frac{1}{2}\left[1 - \frac{R_C}{R_{cp}}\right]\left[1 + \frac{R_C}{R_{cp}} + \frac{R_C^2}{R_{cp}^2}\right] \right\} \qquad (8.1)$$

where the geometric variables are given in the figure and R_C is the radius of the truncation $(R_C = C/\tan \alpha \sim C/\alpha)$. The influence of the truncation can be ignored for most cases, as C is usually quite small in comparison to the radius of the plate; however, when miniature devices are used this can produce a significant effect on the total volume.

The prime variable which minimises the volume is the radius of the plates, as the volume is essentially proportional to the radius cubed. For example, if the radius of the plate is 25 mm and the cone angle 0·1 radians (5·73°) then the sample volume is 3280 μl where the effect of the

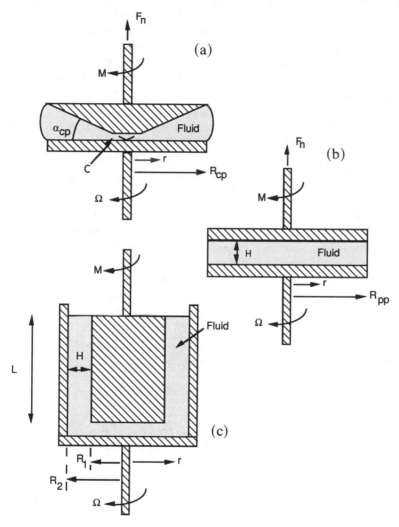

Fig. 8.1. The various torsional geometries: (a) cone and plate, (b) parallel plates and (c) concentric cylinders (the outer cylinder with inner radius R_2 is designated as the cup and the inner cylinder with outer radius R_1 as the bob).

truncation has been ignored (these dimensions are standard for many rheometers). When the radius is decreased to 10 mm, the sample volume becomes 210 μl. The sample volume can be further decreased to 21 μl by using a cone angle of 0·01 radians. This is an acceptable volume to be considered a small sample. If the truncation is 50 μm, the radius 10 mm and the cone angle 0·01 radians, the volume rises to 22 μl (these dimensions are denoted as *small cone dimensions* and the larger dimensions given above with the same truncation are denoted as the *standard cone dimensions*, see Table 8.1). The sample volume has not risen dramatically when the truncation is taken into account, despite the radius of the truncation, R_C, being 5 mm. The volume difference will increase if a smaller plate radius is used.

The shear rate, $\dot{\gamma}_s$, will vary throughout the sample when a truncation is present. The shear rate is given approximately by

$$\dot{\gamma}_s = \frac{\Omega r}{C} \quad \text{for } 0 \le r \le R_C \tag{8.2a}$$

$$\dot{\gamma}_s = \frac{\Omega}{\alpha} \quad \text{for } R_C \le r \le R_{cp} \tag{8.2b}$$

TABLE 8.1

Dimensions and Volume for the Small and Standard Torsional Rheometer Geometries

Geometry	Type	Dimensions[a]	Volume (μl)
Cone and plate	Small	$R_{cp} = 10$ mm $\alpha_{cp} = 0\cdot01$ rad.	21
	Standard	$R_{cp} = 25$ mm $\alpha_{cp} = 0\cdot1$ rad.	3280
Parallel plate	Small	$R_{pp} = 5$ mm $H = 0\cdot25$ mm	20
	Standard	$R_{pp} = 25$ mm $H = 1$ mm	1960
Concentric cylinders	Small	$R_1 = 4$ mm $H = 0\cdot2$ mm $L = 4$ mm	21
	Standard	$R_1 = 20$ mm $H = 1$ mm $L = 20$ mm	2580

[a] See Fig. 8.1 for the definition of the geometric variables.

Taking the linear average over the total radius, R_{cp}, the relative error, E, is determined:

$$E = \frac{\langle \dot{\gamma}_s(C) \rangle - \dot{\gamma}_s(C=0)}{\dot{\gamma}_s(C=0)} = -\frac{1}{2} \frac{C}{\alpha R_{cp}} \tag{8.3}$$

where the angular brackets are used to represent the linear average. E is acceptably small when the standard cone dimensions are used ($E = -1\%$); however, it rises to -25% when the small cone dimensions are assumed. This is an unacceptably large error.

Gap-set errors can also become significant when the cone and plate geometry is miniaturised and occurs by not setting the correct gap between the cone and plate when a truncation is present (see Tanner[5] for a discussion on gap-set errors). The experimental procedure to set the gap is to let the cone and plate achieve thermal equilibrium if they are heated or cooled by a thermal control unit, and move the cone or plate down until they just touch (measured by a normal force read-out or rotating the cone relative to the plate and measuring the torque). The zero is noted, they are moved apart, the sample loaded and allowed to achieve thermal equilibrium; they are then moved together to the truncation height. The minimum error associated with determining the zero gap is approximately ± 1 μm (rheometers are frequently manufactured with linear movements calibrated in 1 μm divisions) and the error in setting the gap to the truncation height is again ± 1 μm. The absolute minimum error in setting the truncation height is ± 2 μm by direct propagation of errors and a probable maximum error is ± 5 μm by assuming a 99% confidence interval on the individual gap measurements. These errors can significantly affect the calculation of the shear rate value throughout the sample, as will be demonstrated below.

When the gap is set at a distance greater than the truncation height, the geometry is termed the extended cone and plate (see Marsh and Pearson[6] and Walters[7]). Call this distance h (the gap error). The gap can also be set smaller than the truncation height, or in other words h is negative. The shear rate is given to a first approximation by

$$\dot{\gamma}_s = \frac{\Omega r}{C+h} \quad \text{for } 0 \leq r \leq R_C \tag{8.4a}$$

$$\dot{\gamma}_s = \frac{\Omega r}{\alpha r + h} \quad \text{for } R_C \leq r \leq R_{cp} \tag{8.4b}$$

If h is equal to zero, the equation simplifies to the true cone-and-plate shear rate (eqn 8.2b; this equation assumes that the cone angle is small, so $\tan \alpha \sim \alpha$. The variation of shear rate in the cone-and-plate geometry has been discussed by Adams and Lodge,[8] and eqn (8.4) with $h = 0$ and eqn (8.2b) do not strictly apply. The above analysis includes the effect of the truncation; however, if a sharp cone is assumed, the following relative error is obtained for the extended cone and plate ($h > 0$) by taking the linear average along the radius,

$$E = \frac{\langle \dot{\gamma}_s(h>0, C=0) \rangle - \dot{\gamma}_s(h=0, C=0)}{\dot{\gamma}_s(h=0, C=0)} = -\frac{h}{\alpha R_p} \ln\left(1 + \frac{\alpha R_p}{h}\right) \quad (8.5)$$

The error is $-7 \cdot 9\%$ and -15% for the small cone dimensions, assuming h is equal to 2 μm and 5 μm, respectively. The standard cone dimensions have errors of $-0 \cdot 6\%$ and $-1 \cdot 2\%$ for the same values of h. One can see that the error becomes quite significant for the small cone dimensions, and combined with the error associated with the truncation may be even larger. Note that this effect produces an average shear rate less than that prescribed.

The error associated with improper gap settings can be elucidated by determining the shear stress, σ_s, for a Newtonian fluid. The torque, M, is measured and is related to the shear stress by

$$M = 2\pi \int_0^{R_{cp}} \sigma_s r^2 \, dr \quad (8.6)$$

Assuming the shear stress is directly proportional to the shear rate, as is true for a Newtonian fluid with viscosity, μ, and that the shear rate is given by eqn (8.4) one arrives at

$$\frac{3}{2} \frac{M}{\pi R_{cp}^3 \mu} = \frac{\sigma_s(C, h)}{\mu}$$

$$= \frac{\Omega}{\alpha} - \frac{\Omega}{\alpha} \left\{ \frac{\frac{1}{4}C + h}{C + h} A^3 + \frac{3}{2}B[1 - A^2] \right.$$

$$\left. - 3B^2[1 - A] + 3B^3 \log\left(\frac{1+B}{A+B}\right) \right\} \quad (8.7)$$

where

$$A = \frac{C}{\alpha R_{cp}}, \qquad B = \frac{h}{\alpha R_{cp}}$$

The quantity σ_s/μ is the apparent shear rate one obtains in the truncated cone and plate, where the gap is set a distance h above ($h>0$) or below ($h<0$) the truncation height, C, for a Newtonian fluid. Equation (8.7) reduces to Ω/α for C and h equal to zero, which is the true shear rate for the ideal cone-and-plate geometry (see eqn (8.4)) and is the true shear rate for any fluid type under these conditions.

The error associated with the shear stress measurement, E_s, is found to be

$$E_s = \frac{\sigma_s(C, h) - \sigma_s(0, 0)}{\sigma_s(0, 0)}$$

$$= -\frac{\frac{1}{4}C + h}{C + h} A^3 - \frac{3}{2}B[1 - A^2] + 3B^2[1 - A] - 3B^3 \log\left(\frac{1+B}{A+B}\right) \quad (8.8)$$

Note that E_s is not zero even for h equal to zero; however, it scales as A^3, which will make the error quite small for the standard geometry yet is -3.1% for the small geometry. This is due to effect of the truncation.

The error for the small geometry, together with results assuming a smaller truncation height, are given in Table 8.2. They indicate that if a smaller truncation height is used ($C=10$ μm) then the error can be much greater than when the larger truncation height is used ($C=50$ μm) if the

TABLE 8.2
A Comparison of the Shear Stress Error Calculated with
Eqn (8.8) for a Newtonian Fluid with the Small Cone-and-
Plate Geometry Assuming Two Truncation Heights

$C=10$ μm		$C=50$ μm	
h (μm)	E_s (%)[a]	h (μm)	E_s (%)[a]
−5	8·3	−5	3·9
−2	3·1	−2	−0·42
0	−0·025	0	−3·1
2	−2·9	2	−5·7
5	−6·9	5	−9·3

[a] The average of the absolute value of the error for $C=10$ μm is 4·2% and for $C=50$ μm is 4·5%.

gap height is set less than C. Interestingly, the average of the absolute error is equivalent between the two. Pritchard[9] has shown that the error in the normal force measurement, F_n, scales as A^2, which will make this error greater than the error associated with the shear-stress measurement.

A summary of the error in the shear-stress measurement for the small and standard geometries is shown in Fig. 8.2. As expected, the error is much greater for the small geometry and can be approximately $\pm 5\%$ for small values of gap error, h, even if a smaller truncation height is used (see also Table 8.2). Equation (8.8) was tested by measuring the error in the shear stress for a 1 Pa s nominal viscosity oil at a shear rate of $10\ s^{-1}$, with a geometry intermediate between the small and large. The results are given in Fig. 8.3. Equation (8.8) is shown to represent the data adequately, giving credence to the above approximate analysis. Thus, one can conclude that the errors associated with miniaturising the cone-and-plate geometry can be significant when the gap is not set properly. Even if the gap is set to the minimum error ($\pm 2\ \mu m$), one can see that decreasing C can exacerbate the problem (see Table 8.2).

Finally, it should be mentioned that it is important that the dimensions of the cone and plate be accurately measured especially when a miniature

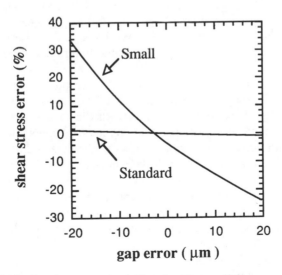

Fig. 8.2. Error in the shear stress calculated with eqn (8.8) assuming the small and standard geometries' dimensions in Table 8.1.

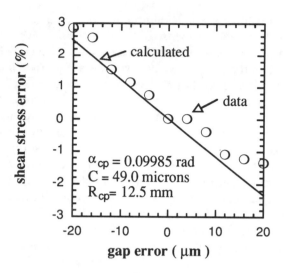

Fig. 8.3. Error in the shear stress calculated with eqn (8.8) compared to data gathered with a cone-and-plate torsional rheometer (the cone-and-plate dimensions are given in the figure). Unpublished results of Mackay.

device is used. The most critical dimensions will be cone angle and truncation height because of their small size, and sophisticated techniques should be employed for their measurement. We have observed errors of up to 2% in the quoted angle from instrument manufacturers in our laboratory (see Walters[7] for a discussion of this).

Other errors which can significantly affect results gathered using this geometry are the shape of the surface, tilt of the cone relative to the plate and eccentricity of the axis of rotation with the axis of the non-rotating member. These effects are discussed by Walters,[7] Adams and Lodge[8] and Pritchard[9] and will certainly cause errors in the measurement if they are present with miniature cone-and-plate devices. The book by Walters[7] gives a detailed account of errors other than those listed above. Most of the above errors will affect the measurement of normal stresses more than the viscosity, as the normal stresses are usually more sensitive to the magnitude of the shear rate than the viscosity (assuming that the fluid is in fact viscoelastic).

The interesting phenomenon of edge fracture was discussed by Hutton[10] in a series of papers. The fracture occurs when the first normal stress difference, N_1, reaches a critical value, N_{1c}, and limits the shear-rate range one can use. Hutton arrived at the following equation which relates N_{1c} to the surface tension, S, and geometrical variables:

$$N_{1c} = \frac{n'S}{R_{cp}\alpha_{cp}} \tag{8.9}$$

where n' is an empirical constant. Note that the denominator in eqn (8.9) is merely the gap at the edge of the cone and plate. Hutton arrived at this equation by an energy analysis which relates the elastic energy to the energy required to form a new surface. He found $n'S$ to be equal to 515 mN/m for a silicon oil, and assuming a surface tension of 21·5 mN/m one finds n' equal to 24. The value was the same for the parallel-plate geometry where the denominator in eqn (8.9) is the gap height indicating that the edge-fracture phenomenon is not geometry-dependent (the parallel-plate geometry is discussed below).

One can calculate the effect of miniaturisation by assuming the shear rate is constant throughout the sample, that the same shear rate is used for the standard and small cone-and-plate geometries and that N_1 is proportional to the shear rate squared (as is true for ideal elastic liquids). From these assumptions one expects the critical shear rate to be a factor of five times higher for the small geometry than the standard. This is a clear advantage for using small samples as this can extend the useful shear-rate range of the cone-and-plate geometry.

Other advantages in using the miniature cone-and-plate geometry are that the shear rate will be essentially constant throughout the sample (neglecting the effect of the truncation) since a small cone angle is used, and inertial effects are smaller. The reader is referred to Walters[7] and Adams and Lodge[8] for a discussion of these effects.

Despite these advantages for using the small geometry, *a miniature cone and plate is not recommended unless great care is taken* because the likely errors in measurement will outweigh the advantages. The results of Benbow and Lamb[1] do demonstrate that this geometry can be used to generate reliable data. They used a geometry with R_{cp} and α_{cp} equal to 5 mm and 0·0524 rad, respectively ($V = 13·7 \, \mu$l). The reason we used a larger plate radius for the calculations above is to ensure that a reasonable torque output is generated when retrofitting a standard rheometer.

8.2.2. Parallel Plates

The parallel plates geometry is shown in Fig. 8.1(b). One plate rotates whilst the other is stationary, and the torque and normal force measured.

The volume contained between the two plates is given by,

$$V = \pi R_p^2 H \tag{8.10}$$

where the geometric variables are defined in the figure. The sample volume is not as strong a function of the plate's radius as for the cone-and-plate geometry. A sample volume of approximately 20 μl can be achieved with this geometry by using a gap of 0·25 mm and a plate radius of 5 mm, or a gap of 0·1 mm and a radius of 7 mm. The former will be used in subsequent calculations and are denoted as *small plate dimensions*. These dimensions together with the *standard dimensions* are listed in Table 8.1. The volume using the standard plate dimensions is 1960 μl. Note that fairly small torque values will be generated with the small parallel plates' assumed radius.

The shear rate in this geometry is a function of the radius measured from the centre, r, and is given by (see Fig. 8.1 for the geometric variables definitions)

$$\dot{\gamma}_s = \frac{\Omega r}{H} \tag{8.11}$$

The main disadvantage of this geometry is that the shear rate is not constant throughout the sample. The shear stress for a fluid is related to the torque by eqn (8.6) with the upper value of the integral equal to R_{pp}. Changing the variable of integration from r to the shear rate and differentiating the integral with respect to the maximum shear rate at R_{pp}, $\dot{\gamma}_s(\text{max})$, one finds

$$\sigma_s(\text{max}) = \frac{2M}{\pi R_{pp}^3} \left\{ \frac{3}{4} + \frac{1}{4} \frac{d\log(M)}{d\log(\dot{\gamma}_s(\text{max}))} \right\} \tag{8.12}$$

where $\sigma_s(\text{max})$ is the shear stress at $\dot{\gamma}_s(\text{max})$. If a Newtonian fluid is used, then the term in front of the braces is the shear stress, since M will be directly proportional to the shear rate (i.e. on a log–log plot the gradient will be one). Assume that the gap has an error, h, from the prescribed distance, H, as was done for the cone and plate geometry above, then one arrives at,

$$\dot{\gamma}_s = \frac{\Omega r}{H + h} \tag{8.13}$$

Using eqns (8.11) and (8.13), the error in the shear rate is easily calculated:

$$E = \frac{\dot{\gamma}_s(H, h) - \dot{\gamma}_s(H, 0)}{\dot{\gamma}_s(H, 0)} = -\frac{h}{H + h} \tag{8.14}$$

The error in the shear rate is fairly small for this geometry. Assuming a gap error of +5 μm, the shear-rate error is −2% and for −5 μm, +2% for the small geometry (±0.5% for the standard). These errors are less than those for the small cone-and-plate geometry (compare to Fig. 8.2 and Table 8.2) and may be acceptable.

Small gaps were thought to influence the viscosity. Burton *et al.*[11] have claimed that the gap distance between parallel plates can substantially influence the viscosity when below 0·25 mm for polystyrene melts. This has been shown by Kalika *et al.*[12] to be an anomalous effect due to leaving untrimmed sample beyond the edge of the plates. This effect has been confirmed in our laboratory, and the viscosity as well as the storage and loss moduli were found to be independent of gap distances from 0·2 to 2·5 mm for a linear low-density polyethylene melt as long as the sample's edge is trimmed flush with the plates. This emphasises the need to trim the polymer and ensure the edge is properly maintained.

Dynamic shear properties, such as the storage and loss moduli, are measured when one plate oscillates relative to the other. When small gaps are used, one must take into account the small yet finite movement of the non-oscillating plate, even when a force rebalance torque head is used.[13] The movement can profoundly affect the torque output, which cannot be used to calculate true dynamic shear properties in this geometry as well as the small cone and plate. In fact, the errors seen by Halley and Mackay[13] were so large that negative values of the loss modulus resulted. Similar effects will be present with the spring-type torque head.

This geometry is less sensitive to gap-set errors than the cone and plate and is recommended for miniaturisation. It is also much easier to fabricate. However, it will be subject to similar errors, such as tilt of the plate, eccentricity of the axis of rotation, etc. The main disadvantage of this geometry is that the shear rate is not constant throughout the sample; however, the true shear stress is easily found from eqn (8.12). Great care should be exercised though, especially in light of the results of Halley and Mackay.

8.2.3. Concentric Cylinders
The concentric cylinder geometry, shown in Fig. 8.1, has one cylinder inside another, where one rotates relative to the other. The sample is placed in the gap between the two. The volume of sample is

$$V = \pi L H \{ H + 2R_1 \} \qquad (8.15)$$

where the geometric variables are given in the figure. The volume is affected by three variables: L, H and R_1, and their influence on the volume will be discussed below.

This geometry is slightly more complicated than the previous two torsional geometries. The shear rate is not homogeneous through the annular gap, and corrections which necessitate that derivatives of the measured shear stress at a given apparent shear rate greater than the first must be taken (described below).[14] The corrections increase in magnitude if the relative gap ratio, H/R_1, which we designate as κ, is relatively large. Thus, a constraint should be used in the design to ensure that κ is small. However, even if κ is small great care should still be used in determining the true shear rate for fluids which exhibit a yield stress.[15]

Also, end effects can influence the measurements, especially at the bottom of the rotating cup shown in Fig. 8.1(c). This is overcome in part by making the bottom of the non-rotating bob a cone which will give essentially the same shear rate as in the annular gap.[16] End effects at the top of the bob can be reduced by filling the liquid to the height of the bob. The shape of the surface should be monitored, as centrifugal forces can push the fluid outwards and elastic forces can cause the rod-climbing effect, both of which reduce the effective area being sheared. End effects can be minimised by making L/H, designated as k, large. This is another constraint in the design.

The volume can be written in dimensionless form in terms of κ and k as

$$\frac{V}{\pi k [R_1]^3} = \kappa^3 + 2\kappa^2 \qquad (8.16)$$

The results from this equation are shown in Fig. 8.4. If k and R_1 are kept constant, then the volume of the sample increases by a factor of 4·1 when κ is increased from 0·05 to 0·1. One can see that the gap ratio has a large effect on the sample volume. The following two constraints can be used for the design of the concentric cylinders geometry:

$$\kappa = \kappa_c \quad \text{(Constraint 1)} \qquad (8.17a)$$

and

$$k = k_c \quad \text{(Constraint 2)} \qquad (8.17b)$$

Where the subscript c represents a constraint and from eqn (8.16) we arrive at,

$$R_1 = \left\{ \frac{V}{\pi k_c [\kappa_c^3 + 2\kappa_c^2]} \right\}^{\frac{1}{3}} \qquad (8.18)$$

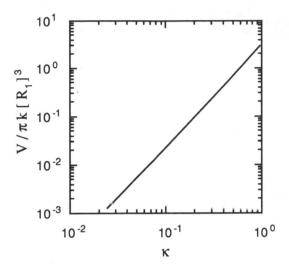

Fig. 8.4. The relation between the dimensionless volume and dimensionless gap for the concentric cylinder geometry; k is L/H and κ is H/R_1.

κ_c and k_c are assumed to be 0·05 and 20, respectively, making R_1, H and L equal to 4 mm, 0·2 mm and 4 mm, respectively, for a volume of 21 µl. These values should reduce errors in the shear rate and shear stress while maintaining a reasonable size. The standard dimensions are listed in Table 8.1 and require a volume of 2580 µl which is considerably larger than the small geometry's volume (the same values κ_c and k_c were assumed).

The shear stress at the bob, σ_{s1}, in this geometry is given by

$$\sigma_{s1} = \frac{M}{2\pi LR_1^2} \qquad (8.19a)$$

which is determined by dividing the torque by the surface area of the bob and the lever arm, R_1. The shear stress at the inside of the cup, σ_{s2}, is,

$$\sigma_{s2} = \sigma_{s1} \left\{ \frac{1}{1+\kappa} \right\}^2 \qquad (8.19b)$$

Even with κ equal to 0·05 the shear stress at the outer cylinder is 91% of that at the inner. Thus, the shear stress is not homogeneous throughout the sample, which can produce anomalous results, particularly with yield-stress fluids.

The shear rate is found, from a force balance and assuming appropriate kinematics, to yield[6,14]

$$\Omega = \frac{1}{2} \int_{\tau_{s2}}^{\tau_{s1}} \dot{\gamma}_s \frac{d\sigma_s}{\sigma_s} \tag{8.20}$$

The shear rate is not constant throughout the sample as can be inferred from eqn (8.19b). The true shear rate at the bob (where one measures the shear stress), $\dot{\gamma}_{s1}$, is easily found for a Newtonian fluid:

$$\dot{\gamma}_{s1} = \frac{2\Omega}{1 - \varepsilon^2} \tag{8.21}$$

where

$$\varepsilon = \frac{1}{1 + \kappa} \tag{8.22}$$

or for a power law fluid (the shear stress is proportional to the shear rate raised to a power n; this relation is assumed true for all shear rates present in the annular gap)

$$\dot{\gamma}_{s1} = \frac{2\Omega}{n[1 - \varepsilon^{2/n}]} \tag{8.23}$$

The ratio of the shear rate for a Newtonian fluid to that of a power law fluid, assuming n is 0·5, is 0·954 for both the small and standard geometries. Thus, a 5% error would be present if eqn (8.21) were used rather than eqn (8.23). The error is the same for the two geometries as κ is the same for both; however, the small geometry is more sensitive to alignment errors, which will be addressed below.

Zimm and Crothers[17] and McCutchen[18] have developed small concentric-cylinder rheometers. The design of Zimm and Crothers required 3–10 ml of fluid which is too large to be considered a small volume; however, their design was self-centring. The bob floated in the cup and centred itself through surface-tension forces. This is an advantage unique to this type of concentric cylinder design. The glass bob had a piece of steel mounted in the bottom with ballast to achieve the correct vertical height in the glass cup. A rotating magnetic field was applied below the cup concentric with the centre-line of the rotation. The rotation rate of the bob was measured and compared to a standard fluid, for example, the solvent, and the relative viscosity was determined. The accuracy in measurement was excellent with this device, being within 0·5% of measurements with a capillary device. Note that this design operates at constant shear stress through the applied magnetic field and that there is no mechanical friction present. This self-centring instrument could possibly be miniaturised.

McCutchen required a sample volume of 200 µl; however, he commented that his design could be used for smaller sample volumes. He used a similar design to Zimm and Crothers except that the bob had a permanent magnet in it and the cup rotated. An external electromagnet was used to keep the bob from rotating, with the supplied current proportional to the shear stress. The shear rate was calculated by the rate of rotation of the cup. This instrument operates at constant rate rather than constant stress.

McCutchen as well as Walters[7] discuss alignment errors in this geometry in detail such as tilt of either member and eccentricity of the rotation axis to the centre-line of the instrument. We discuss the influence of the latter alignment error below where the rotation axis is displaced by an amount a from the instrument centre-line for a Newtonian fluid.[19] Defining the relative amount of eccentricity as $\delta(\equiv a/H)$ one can determine the error in the shear stress, $\sigma_{s1}(\delta)$, to that where no eccentricity is present, $\sigma_{s1}(0)$,

$$E_s = \frac{\sigma_{s1}(\delta) - \sigma_{s1}(0)}{\sigma_{s1}(0)} = \frac{1 + 2\delta^2}{1 + \frac{1}{2}\delta^2} \frac{1}{\sqrt{1 - \delta^2}} - 1 \qquad (8.24a)$$

or

$$E_s \approx 2\delta^2 + \frac{3}{4}\delta^4 + \cdots \qquad (8.24b)$$

where E_s is the error in the shear stress measurement. If one assumes an eccentricity of 5 µm, E_s is only 0·125% for the small geometry and 0·005% for the standard. These are negligible errors; however, if the eccentricity is increased and the gap decreased then this error can become appreciable. Also, if the fluid is viscoelastic then the error can increase as the flow is partly converging, thereby increasing the stresses and hence the torque.[19]

This geometry does seem appealing for a miniature rheometer. Despite the calculations presented above, the errors can become significant if alignment is not almost perfect (i.e. 5 µm is a fairly good eccentricity alignment for this geometry). Also, sample loading is difficult and entrapped air bubbles may be difficult to remove.

8.3. FALLING-BALL RHEOMETER

Many researchers have used the falling-ball rheometer in the past. It consists of a ball which falls either by gravity or an external applied field in a tube of diameter D and length L. Two research groups at Duke University[20] and Johannes Kepler University[21] have recently investigated the falling-ball rheometer. Both groups use a magnetic field to move the sphere and only differ in the detection system. The group at Duke University use ultrasound echo to determine the motion of the sphere, whilst at Johannes Kepler University they either shine a laser light through a window and trace the position of the sphere by a photodiode array, or use an induction coil whose impedance is changed as the ball passes through the coil. The former group have concentrated on measurement of dynamic shear properties by oscillating the sphere and the latter the steady shear viscosity by making the sphere move steadily down the tube.

The volume of sample required is easily calculated as

$$V = \frac{\pi D^2}{4} L \qquad (8.25)$$

A sample volume of 20 µl is achieved by using a tube with a diameter of 1·6 mm and length 5 mm (or really the length of sample in the tube).[20(a)] A large diameter is needed as the sphere must be relatively large (~ 1 mm) so the magnetic field can move it in viscous fluids; smaller spheres are difficult to detect with either detection system.

The magnetic field strength, F_m, is given by

$$F_m = \rho_s V_s \chi H(x) \frac{\partial H(x)}{\partial x} \qquad (8.26)$$

where ρ_s is the density of the sphere, V_s, the sphere volume, χ, the magnetic susceptibility of the sphere and $H(x)$, the magnetic field strength at any position, x, along the tube. The influence of gravity and buoyancy forces will not be considered, but is easily included in the subsequent analysis if need be. Note that, as the above equation implies, the magnetic field is not homogeneous along the tube length and a suitable technique must be used to find its value as well as its derivative at any given position. Under equilibrium conditions, F_m is equal to the hydrodynamic force, F_h, on the sphere where

$$F_h = 3\pi \, d\eta v K(d/D, n, We) \qquad (8.27)$$

d is the sphere diameter, η, the fluid's viscosity, v, the terminal velocity of the sphere and K, a correction factor to account for the fact that the hydrodynamic force is influenced by the tube (d/D),[22] shear thinning effects (n, which is the power law index and equals unity for a Newtonian fluid)[23] and the fluid elasticity ($We \equiv$ Weissenberg Number $= 2\lambda v/d$; λ is a suitable relaxation time and is equal to zero for a Newtonian fluid).[24] K equals unity when the above effects are negligible. Setting the two above equations equal, one arrives at

$$\eta = \frac{\rho_s d^2 \chi H(x)}{18vK(d/D, n, We)} \frac{\partial H(x)}{\partial x} \tag{8.28}$$

The correction factor, K, deserves some comment at this point. Happel and Brenner[22] give the following equation for a Newtonian fluid:

$$K(d/D, 1, 0)^{-1} = 1 - 2 \cdot 10443 \left[\frac{d}{D}\right] + 2 \cdot 08877 \left[\frac{d}{D}\right]^3 - 0 \cdot 94813 \left[\frac{d}{D}\right]^5$$

$$- 1 \cdot 372 \left[\frac{d}{D}\right]^6 + 3 \cdot 87 \left[\frac{d}{D}\right]^8 - 4 \cdot 19 \left[\frac{d}{D}\right]^{10} \tag{8.29}$$

If one assumes that the sphere diameter is 1 mm and the tube diameter 1·6 mm, then the correction factor is 13·43, which is quite large. All the terms in the above expansion significantly contribute to the final answer, suggesting that more terms are required if it is to be used for this value of d/D and that perhaps a more accurate technique would be to determine the correction factor experimentally.

Dazhi and Tanner[23] used a finite element numerical simulation to determine the influence of shear thinning ($n \leq 1$, see also Sugeng and Tanner[25]). They present their results as

$$K(d/D, n, 0)^{-1} = 1 - \alpha(n) \left[\frac{d}{D}\right] \quad \text{for} \quad \left[\frac{d}{D}\right] \leq 0 \cdot 05 \tag{8.30}$$

where $\alpha(n)$ can be adequately represented by

$$\alpha(n) = -0 \cdot 568n + 2 \cdot 672n^3 \quad \text{for} \quad \left[\frac{d}{D}\right] \leq 0 \cdot 05 \tag{8.31}$$

Note that this equation is valid only for relatively small values of the ratio of the sphere to tube diameter. However, it does give an indication of the influence of shear thinning. If d/D is 0·05 and n is 0·5, then the correction factor is 1.003 compared to 1.118 for a Newtonian fluid. This indicates that shear viscosity will be underestimated by eqn (8.28) if the Newtonian wall correction is used.

Chhabra et al.[24] have determined the influence of elasticity on the hydrodynamic force by using highly elastic, constant-viscosity fluids.[26] These fluids do not exhibit shear thinning and eliminate confusion as to whether shear thinning or elasticity is producing an effect in a given experiment. The above researchers unfortunately do not consider wall effects. Their results can be summarised as

$$K(0, 1, We) = 1 \quad \text{for } We \leq 0.1 \tag{8.32a}$$

$$K(0, 1, We) = 0.692 - 0.308 \log_{10}(We) \quad \text{for } 0.1 \leq We \leq 0.7 \tag{8.32b}$$

$$K(0, 1, We) = 0.74 \quad \text{for } 0.7 \leq We \leq 2.02 \tag{8.32c}$$

Even if the viscosity is constant, elasticity can affect the hydrodynamic force and an error of up to 26% can result in the measured viscosity. Consider, for example, a polymer melt where a 1-mm diameter sphere in an infinitely wide tube travels at 5×10^{-3} mm/s; then the relaxation time needs to be 10 s for elastic effects to influence the hydrodynamic force ($We = 0.1$). This magnitude of elasticity is present in many polymer melts. Note that the Reynolds number ($\rho v d / \eta$, where ρ is the fluid's density) is of the order 10^{-10} for these flow conditions, assuming a typical viscosity. Sugeng and Tanner used numerical simulation to find similar values for the correction factor when wall effects are taken into account and found that shear thinning together with elasticity decreases the correction factor to a greater extent ($K < 0.74$).

Tran-Son-Tay et al.[20(c)] have addressed the influence of inertia and wall effects in dynamic shear where the sphere is forced to oscillate along the symmetry axis of the tube. They find that both of these greatly affect the results.

One should note that the shear rate and shear stress are not constant over the surface of the sphere, which can induce errors in measurement. The needle geometry could be used to minimise this effect.[27] Thus, one can conclude that this geometry does require care in interpretation of the results. However, it is an extremely useful geometry and can be used to determine the effect of pressure on the viscosity and to watch cure reactions with relative ease.[21]

8.4. CAPILLARY RHEOMETER

Benbow and Lamb[1] have used a miniature capillary rheometer to determine the viscosity of polymer melts; however, they required approximately 10^3 µl of sample. An interesting capillary rheometer design for biological materials was developed by Philippoff *et al.*[28] and was used to determine the fluid's viscosity and the recoverable shear strain, S_R, which is a measure of the elasticity. Their apparatus requires sample volumes of approximately 10 µl. A schematic of the device is shown in Fig. 8.5. The entire flow section was placed in a microscope and viewed under 100× magnification with a eyepiece micrometer in place.

The experimental procedure was to pull the sample a distance L into the larger capillary by vacuum and to let it relax. A slight constant vacuum (or strictly a pressure drop) was applied, the steady state velocity determined and the meniscus shape photographed. The meniscus shape was needed for S_R. The shear stress at the wall, σ_w, and the apparent wall shear rate, $\dot{\gamma}_{aw}$, were found by

$$\sigma_w = \frac{d\Delta P}{4l} \tag{8.33}$$

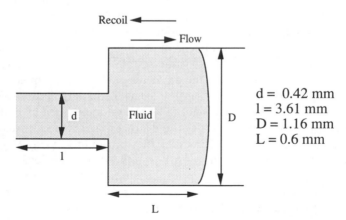

Fig. 8.5. Schematic of the miniature capillary rheometer of Philippoff *et al.* for measuring the shear viscosity and recoverable shear strain of biological materials. Note the shape of the meniscus is not flat in the larger capillary and its shape will be different during flow and under static conditions.

and

$$\dot{\gamma}_{aw} = \frac{32Q}{\pi d^3} \tag{8.34}$$

where the geometrical variables are given in the figure, ΔP is the applied vacuum and Q the volumetric flow rate. The true shear rate at the wall, $\dot{\gamma}_w$, is found by the Rabinowitsch correction[7]

$$\dot{\gamma}_w = \dot{\gamma}_{aw} \left\{ \frac{3}{4} + \frac{1}{4} \frac{d \log(\dot{\gamma}_{aw})}{d \log(\sigma_w)} \right\} \tag{8.35}$$

The recoverable shear strain was found by the volume regained when the vacuum was turned off, and is related to the memory of the fluid. The meniscus retracted by a value X, which is related to S_R by

$$S_R = \frac{8kX}{d} \left\{ \frac{D}{d} \right\}^2 \tag{8.36}$$

where k is a correction factor to account for the difference in the meniscus shape during flow and at rest, and may be determined by comparing photographs. In other words, the different meniscus shapes will produce a volume difference that must be accounted for when the linear distance, X, is measured.

The results they obtained were in good agreement with other techniques for a solution of 3% polyisobutylene in decalin for both the viscosity and S_R. Thus, this technique appears valid despite its small dimensions.

One should be aware that if the diameter of the larger capillary, D, is comparable to the smaller, d, then the shear in the larger capillary may influence the results. However, since the larger diameter used by Philippoff *et al.* was approximately a factor of three larger than the smaller diameter, the apparent shear rate is 27 times smaller in the larger capillary. This should not drastically affect the overall results. Also, if the device is made too small, capillarity effects may produce anomalous results, particularly for the measurement of S_R. This does not seem to be a problem since they show good agreement between various techniques.

Finally, small capillary diameters can produce a variety of unwanted effects such as: the influence of boundaries on small molecular weight liquid crystalline polymers;[29] and apparent slip of suspensions,[30] emulsions,[31] gels,[32] polymer solutions,[33] polymer melts,[34] and liquid crystalline polymers.[35] These references are only representative of the diameter effects in capillary rheometry; use of various diameters and length-to-diameter ratios are recommended to ensure the data are not affected by slip and entrance or exit effects, respectively.

The clever design of Philippoff *et al.* allows one to determine the shear viscosity and recoverable shear strain of a small amount of fluid, particularly biological fluids. The technique is somewhat labour-intensive, as is capillary rheometry in general; however, with a video recorder and modern digitisation techniques, this technique could be easily used.

8.5. SURFACE-FORCES RHEOMETER

The surface-forces rheometer is a modification of the surface-forces apparatus discussed by Israelachvili and Tabor.[4] The original device consisted of two crossed mica cylinders driven together to extremely small distances, of the order of nanometres. The force was measured by a cantilever spring and the separation distance by observing interference fringes (Fringes of Equal Chromatic Order, FECO). The crossed cylinders are hydrodynamically equivalent to a hypothetical sphere of radius R, approaching a flat. Chan and Horn[36] have discussed the hydrodynamics of the geometry in detail. The volume of fluid required is given by

$$V = \pi h^2 \left\{ R - \tfrac{2}{3}h + \left[2\frac{R}{h} - 1 \right] D \right\} \tag{8.37}$$

where the geometric variables are given in Fig. 8.6(a) and h is the immersion depth of the hypothetical sphere in the fluid. The third term in the curly brackets can be neglected for most separation distances. Assuming R is equal to 1 cm and h, 1 mm, the required volume is approximately 30 μl. Of course, smaller volumes will be required if h is made smaller, since the volume is essentially proportional to h^2; so this technique can be classified as suitable for small samples.

The surface-forces rheometer operates on exactly the same principle except that one of the cylinders is allowed to vibrate vertically about a mean separation distance , $\bar{D}$. A schematic of the apparatus is shown in Fig. 8.6(a). This geometry was used by Israelachvili and coworkers[3,37] and van Alsten and Granick[38] (these researchers used a sliding oscillatory geometry where the upper member moves horizontally rather than vertically; the force measurement was suitably modified). A group in France[39] have used a sphere-and-flat geometry rather than the crossed cylinders

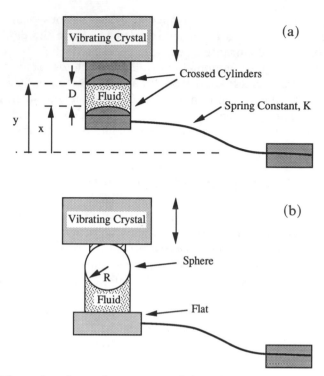

Fig. 8.6. The surface-forces rheometer consisting of crossed cylinders (a) and a sphere and flat (b). The hydrodynamics of the crossed cylinders is equivalent to a sphere and flat.[36]

(see Fig. 8.6(b)). Both geometries operate in the same manner and the analysis given below is essentially the same for each.

Israelachvili gave the following equation to determine the viscosity, η, of a liquid where the crystal vibrates at an amplitude A, and frequency f (Hertz):

$$\eta = \frac{K\bar{D}}{12\pi^2 R^2 f}\left\{\left[\frac{A_D}{A}\right]^2 - 1\right\}^{\frac{1}{2}} \quad \text{for } f \to 0 \tag{8.38}$$

where A_D is the amplitude of the intersurface separation which is measured by the FECO. Other equations should be employed when larger frequencies are used.[37] Equation (8.38) shows that the determination of A_D is critical and the FECO must be accurately measured. Two techniques are used[37(b)] depending on the frequency.

The operation of this rheometer is non-trivial and great care should be exercised.[40] Indeed, the relation of the total force measured is a complicated function of the separation distance.[41] Nevertheless, interesting results can be obtained regarding the effect of intersurface separation distance on the rheological properties. Chan and Horn[36] have found that the viscosity of three non-polar Newtonian liquids shows an increase when the separation distance is approximately 5 nm. Their results were determined by driving the lower cylinder towards the upper in a continuous manner. This is contrary to the results of Israelachvili[3] who found the viscosities for similar fluids were independent of gap separation. The shear rates used by Israelachvili were approximately a factor of 6 less than those of Chan and Horn and this was hypothesised to account for the difference. Also, one should note that the shear fields are not the same; one constantly increases while the other oscillates with time.

It is not the purpose of this Chapter to discuss separation distance effects on the rheological properties; however, the results of Montfort et al.[39(b)] do deserve some comment, especially in light of the results of Burton et al.[11] They find that the rheological properties of a concentrated solution of polybutadiene in a hydrocarbon oil (26% by volume, molecular weight of 43 000 Daltons) are independent of gap for separation distances of the order of microns. The results compare well to data gathered with a standard cone-and-plate geometry. This is contrary to the results of Burton et al., who show deviations when gaps are below 250 μm, which reinforces the conclusions of Kalika et al.[12] that these results were subject to measurement errors. Thus, it can be concluded that the rheological properties of polymer solutions and possibly polymer melts are not affected by gap distances of the order of microns. However, one should note that the highest frequency used by Montfort et al. was 10^3 rad/s and the average relaxation time was 1·6 ms, so the product of the two is approximately unity. It would be interesting to perform experiments when the frequency is well above the longest relaxation mode of the polymer at gaps of the order of the radius of gyration. A gap effect may become evident.

In addition, the results of van Alsten and Granick show interesting gap effects. They used very large strain amplitudes ($A_H/\bar{D} \sim 10$, where A_H is the horizontal vibrational amplitude) and found that the signal amplitude decreased with time when the gap was small. This is remarkably similar to the predictions and experimental data of Hatzikiriakos and Dealy[42] using polymer melts in a sliding-plate rheometer. Their gaps were relatively large (~ 230 μm) as were their strain amplitudes (~ 20); however, they found that the stress amplitude decreased by a factor of approximately 3 with time and changed from being sinusoidal to non-sinusoidal. They interpreted their results in terms of a sophisticated slip model. Close inspection of the data of van Alsten and Granick shows that their response (i.e. stress) waveforms are slightly non-sinusoidal and decrease by the same order. It could be that these experiments are demonstrating similar effects. More experiments are needed to delineate the effects of surfaces on rheological properties before firm conclusions can be made.

In conclusion, this type of rheometer is an important development in measuring the influence of gap size on rheological properties. Hopefully future results from this rheometer can be used to understand the complicated behaviour of polymers near surfaces under shear.

8.6. PRONG RHEOMETER

The prong rheometer was developed by Mackay and Cathey[43] and is most useful for measuring the dynamic shear properties of complex fluids. A schematic of the geometry is shown in Fig. 8.7. The lower prong rotates relative to the upper one, where the torque is measured. The fluid is confined between the two square, flat plates and held in place by surface tension forces. The volume of sample required is,

$$V = NblH \tag{8.39}$$

where N is the number of plate pairs used (two plate pairs are shown in the figure). Typical dimensions of the prong are 5 mm, 5 mm and 0·5 mm for b, l and H, respectively. The volume needed is $N \times 12·5$ μl or 25 μl for the apparatus shown; however, as long as the torque generated is sufficient, only one arm of the prong can be used (i.e. N is unity). The geometry has been successfully used for a polymer melt with b and l equal to 3 mm and H equal to 0·250 mm, so a total volume of 4·5 μl was needed.[43] The prong geometry approximates the sliding plate[44,45] and will exactly mimic sliding plates in the limit of the lever arms, R_S and R_B, approaching infinity.

Mackay and Cathey discussed the technique to 'trick' a torsional rheometer, so standard software may be used. The shear strain and rate are assumed constant throughout the sample; of course, this is not exactly true and is valid when b and l are much smaller in magnitude than R_S and R_B. This is discussed below. Assuming the above is true, the prong geometry can be related to any standard torsional geometry where the

Side View

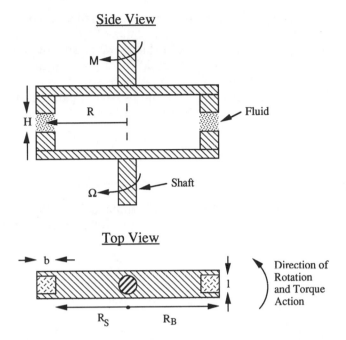

Top View

Fig. 8.7. The prong geometry for a standard torsional rheometer. The fluid is placed between two square, flat plates and one prong rotates relative to the other. The torque is measured and converted to a shear stress. Circular, flat plates can also be used.

shear rate is constant, such as the cone and plate or narrow-gap concentric cylinders. The cone-and-plate geometry was chosen and a 'phantom' plate radius and cone angle were calculated.

The shear rate in the prong geometry at any radial position, R, is

$$\dot{\gamma}_s = \frac{R\Omega}{H} \tag{8.40}$$

Setting this equation equal to eqn (8.2(b)) one arrives at the phantom cone angle, α_p,

$$\alpha_p = \frac{H}{R} \tag{8.41}$$

α_p depends upon the choice of R. Mackay and Cathey discussed this in detail and found that the linear average of R_S and R_B (i.e. $[R_S + R_B]/2$)

gave accurate results. Note their prong geometry had an R_S of 11·00 mm
and R_B of 16·00 mm, so the linear average was 13·50 mm. The shear rate
will vary by approximately 20% from this average value over the prong
area. This error can be reduced by designing a prong with larger radii
relative to the plate dimensions.

The phantom cone radius, R_p, is found by equating the shear stress in
the cone-and-plate geometry to the prong. The shear stress in the prong
geometry is

$$\sigma_s = \frac{F_s}{A} \tag{8.42}$$

where F_s is the shear force at any radial position and A is the area of one
plate ($A = lb$ in Fig. 8.7). The torque is related to the shear force through

$$M = NF_sR \tag{8.43}$$

Equating the shear stress given in eqn (8.7) to that in eqns (8.42) and
(8.43), and assuming F_s is constant, one arrives at

$$R_p = \left\{ \frac{3}{2\pi} NRA \right\}^{\frac{1}{3}} \tag{8.44}$$

Using the typical dimensions given in Mackay and Cathey: R, l, b and H
equal to 13·50 mm, 5 mm, 5 mm and 0·5 mm, respectively, the phantom
cone angle is 0·0370 rad and the phantom cone radius is 6·856 mm.

These values of the phantom cone angle and radius were used with a
torsional rheometer and the dynamic storage and loss modulus determined with the prong geometry were compared to those measured with
a standard cone-and-plate or parallel-plate geometry.[43] The results were
very accurate for a variety of fluids; a polymer melt, a standard silicon
oil and a paint base; as long as a gap of 0·5 mm was used and surface
tension forces were noted. The reason for using a gap of 0·5 mm is not
entirely clear at this point; however, this gap displayed a clear maximum
in the dynamic properties at a given frequency. The maximum was identical to the dynamic properties measured with the cone and plate. This
allowed the proper gap to be selected.

The effect of surface tension forces is easily accounted for. Following
Laun and Meissner[44] the change in energy, ΔW, in deforming the free
surface of fluid contained between two arbitrary, flat surfaces (see

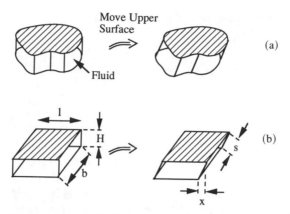

Fig. 8.8. The deformation of the free surface area between two arbitrary, flat surfaces (a) and two flat plates (b). When the upper surface moves relative to the lower a larger surface area is exposed to the surrounding atmosphere.

Fig. 8.8) when the upper is moved a distance Δx is

$$\Delta W = F_S \Delta x = S \Delta A_f \tag{8.45}$$

where F_S is the force due to surface tension, S, the surface tension and ΔA_f, the change in free area. Rearranging the above equation and taking the limit as Δx approaches zero one finds

$$F_S = S \frac{dA_f}{dx} \tag{8.46}$$

This equation applies to any free surface. Consider the two flat plates in Fig. 8.8(b), which is the prong geometry. The free area is,

$$A_f = 2Hl + 2bs = 2Hl + 2bH \sqrt{1 + \left[\frac{x}{H}\right]^2} \tag{8.47}$$

Taking the derivative of eqn (8.47) according to eqn (8.46) one finds,

$$F_S = 2bS \frac{\gamma_s}{\sqrt{1 + [\gamma_s]^2}} \tag{8.48}$$

where, γ_s is the macroscopic shear strain (x/H).

The surface tension of the fluid creates a shear stress contribution, $\sigma_{s,S}$, in addition to the shear stress of the fluid, $\sigma_{s,F}$. The total shear stress,

σ_S, is the sum of these two contributions. From eqns (8.7), (8.43) and (8.44) the total shear stress is

$$\sigma_s = \sigma_{s,F} + \frac{2S}{l} \frac{\gamma_s}{\sqrt{1+[\gamma_s]^2}} \tag{8.49}$$

Thus, the shear stress is increased by the amount given by the second term on the right-hand side of the above equation. Eqn (8.49) differs from eqn (9b) of Mackay and Cathey by a factor of two and is the result of an error in the torque balance. The surface tension found by Mackay and Cathey in a dynamic test for a silicon oil will be a factor of two smaller. The surface tension of the silicon oil was measured by the Wilhelmy plate method which was also found to be in error for this viscous fluid whose viscosity was approximately 100 Pa s. This error was the result of unfortunate edge effects exacerbated by the viscosity of the fluid. The two techniques agree well after these errors were accounted for. It has been subsequently found that if circular, flat disks are used rather than the flat plates shown in Fig. 8.8(b) the measured surface tension is more accurate. This will be presented in a future publication. Laun and Meissner[44] found that the surface tension of a polymer melt was accurately determined during creep recovery when flat plates were used.

The purpose of the above analysis is to demonstrate that the shear stress will be influenced by surface-tension forces if the fluid stress is small in comparison. The dynamic storage modulus at small strain amplitudes will be increased by a constant factor which can be seen by inspection of eqn (8.49). Also, the shear stress will be increased during steady shear; however, the amount by which it is increased will change with time as the shear strain is not constant. Finally, the relaxation modulus will be increased by a constant amount when a step strain is applied. These three tests can be used to determine the surface tension or conversely to indicate the likely error involved in each test.

Since the dynamic modulus is fairly large for polymer melts, this quantity will be subject to small errors. The prong geometry allows testing of a single polymer pellet. One can perform tests on single pellets to look for variations in polymer quality with this geometry. Also, a sample does not need to be melted and formed into a disk prior to testing, which eliminates one thermal cycle. Finally, true, large-amplitude dynamic moduli of suspensions and gels can be measured with this geometry.[43,46] Large amplitudes for these systems are frequently from 1 to 10% strain, and the prong geometry can be used in this range of strains.

Errors in using this geometry are similar to those encountered with the parallel plate. The gap used in the prong geometry is of the order of 0·5 mm so an error of 5 μm will not greatly affect the results. Alignment should be carefully performed, as is true with any geometry. However, the overall effect of alignment errors is reasonably small with this geometry. Finally, it should be noted that relatively small strains should be used. If the strain is too great, then the sample can 'rip' off the surface of the plates, which has been observed in our laboratory. Strains less than 50% can be used with most fluids.

This geometry is useful for measuring the rheological properties of polymer melts, polymer solutions and suspensions. It eliminates a thermal cycle for polymer melts as it requires only a small amount of sample and a single pellet will suffice. Suspensions are easily loaded and air bubbles can be minimised. Some care should be exercised and the effect of surface tension accounted for, particularly for less elastic and viscous fluids.

8.7. OTHER RHEOMETERS

Dintenfass[47] has measured the viscosity of biological fluids using a variety of torsional rheometer geometries. These geometries differ only slightly from those torsional geometries described above. The volume of fluid required is approximately a factor of 10 greater than our classification of small; however, through further miniaturisation, the volume could be reduced.

Miles and coworkers[48] used a geometry similar to the sliding plate[44,45] to measure the dynamic shear properties of a variety of fluids. Unfortunately, Miles does not give the dimensions of the apparatus or volume of fluid required; it is assumed that the volume is small. The apparatus is capable of achieving very high frequencies approaching 10^4 rad/s. Because he used such high frequencies, he was able to measure the plateau modulus of fluids like pentachlorobiphenyl which had a value of approximately 10^9 Pa for temperatures below $-20°C$.

An interesting geometry for measuring the dynamic shear properties of elastomers uses a tuning fork.[49,50] A small amount of sample was placed on each of the two prongs of the fork and a bar was carefully placed on the samples. The ends of the prongs were placed into harmonic motion by electromagnetic telephone receivers. The sample was thus sheared between the prongs and the bar. The signal was measured at the shaft of the fork. The change in frequency and decrement were determined by

monitoring the signal which was converted into the dynamic shear properties (i.e. the loss and storage modulus). This apparatus was capable of
frequencies in excess of 10^4 rad/s. Rorden and Grieco[49] also describe
a horizontal pendulum for measuring the low-frequency dynamic shear
properties.

Finally, a technique for measuring the elongational properties, or really
the 'spinability' of biological fluids, is mentioned.[51] This technique
required only 20 μl of sample. The sample was placed in a small hole
drilled in the measurement cell, which allowed 6 μl to overflow. A clip
was used to grasp the sample and was moved upwards at a constant rate,
thereby stretching the fibre of fluid. The current was monitored between
the clip and measurement cell and the height measured when the current
disappeared (i.e. the fibre of fluid broke). The height in mm was called
the 'spinability'. Various fluids were tested and the results were fairly
reproducible.

8.8. CONCLUDING REMARKS

A variety of rheometers have been discussed which can be used to measure
the rheological properties of small samples. All the rheometers have draw-
backs; this is due to the size of the samples. However, all the geometries
mentioned could be used for accurate measurement of the rheological
properties provided good instrumentation, great care and patience are
used. The only area of research to date where a concerted effort has been
made to measure the rheological properties of small samples is in the
biological sciences. This is by need and not choice since only small
amounts of sample are available. With the increase of research activity in
the biological sciences, the need to measure the true rheological properties
of small samples is certain to increase. Finally, it is expected that in this
age of miniaturisation, small rheometers will be required for small samples
either because the amount of sample available is small, such as in the
biological sciences, or one wishes to determine the effect of sample size
on its rheological properties.

ACKNOWLEDGEMENTS

I thank the Australian Research Council for continued funding. Also, I
would like to thank my students: Peter Farrington, Peter Halley and
Tony Wilson for not believing that the prong rheometer would work until
they each tried it themselves.

NOTE ADDED IN PROOF

Recently, Denn, in a personal communication, has indicated that the results of Burton *et al.*[11] may not be in error. We have used small gaps with a parallel plate geometry on polydisperse polyethylene and saw no effect within experimental error. This subject certainly deserves further attention as it may indicate that anomolous structures develop in polymer melts.

REFERENCES

1. J. J. Benbow and P. Lamb, *J. Sci. Inst.*, 1964, **41**, 203.
2. L. Dintenfass, *Rheology of Blood in Diagnostic and Preventative Medicine*, Butterworths, London, 1976.
3. J. N. Israelachvili, *J. Coll. Int. Sci.*, 1986, **110**, 263.
4. J. N. Israelachvili and D. Tabor, *Proc. Roy. Soc. A*, 1972, **331**, 19.
5. R. I. Tanner, *Engineering Rheology*, Oxford University Press, Oxford, 1988.
6. B. D. Marsh and J. R. A. Pearson, *Rheol. Acta*, 1968, **7**, 326.
7. K. Walters, *Rheometry*, Chapman and Hall, London, 1975.
8. N. Adams and A. S. Lodge, *Phil. Trans. Roy. Soc. Lond. A*, 1964, **256**, 149.
9. W. G. Pritchard, *Phil. Trans. Roy. Soc. Lond. A*, 1971, **270**, 507.
10. J. F. Hutton, *Nature*, 1963, **200**, 646; *Proc. Roy. Soc. A*, 1965, **287**, 222; *Rheol. Acta*, 1969, **8**, 54; in *The Rheology of Lubricants*, Davenport, T. C. (Ed.), Applied Science Publications, Barking, 1973, p. 108.
11. R. H. Burton, M. J. Folkes, K. A. Narh and A. Keller, *J. Mat. Sci.*, 1983, **18**, 315.
12. D. S. Kalika, L. Nuel and M. M. Denn, *J. Rheol.*, 1989, **33**, 1059.
13. P. J. Halley and M. E. Mackay, *J. Rheol.*, 1991, **35**, 1609.
14. (a) I. M. Krieger and S. H. Maron, *J. Appl. Phys.*, 1953, **23**, 147; (b) I. M. Krieger and H. Elrod, *J. Appl. Phys.*, 1953, **24**, 134; (c) I. M. Krieger and S. H. Maron, *J. Appl. Phys.*, 1954, **25**, 72; (d) I. M. Krieger, *Trans. Soc. Rheol.*, 1968, **12**, 5; (e) T. M. T. Yang and I. M. Krieger, *J. Rheol.*, 1978, **22**, 413.
15. R. Darby, *J. Rheol.*, 1985, **29**, 369.
16. M. Mooney and R. H. Ewart, *Physics*, 1934, **5**, 350.
17. B. H. Zimm and D. M. Crothers, *Nat. Acad. Sci. Proc.*, 1962, **48**, 905.
18. C. W. McCutchen, *Biorheology*, 1974, **11**, 265.
19. R. B. Bird, R. C. Armstrong and O. Hassager, *Dynamics of Polymeric Liquids*, Vol. 1, 2nd Edn, John Wiley and Sons, Brisbane, 1987.
20. (a) R. Tran-Son-Tay, B. B. Beaty, D. N. Acker and R. M. Hochmuth, *Rev. Sci. Inst.*, 1988, **59**, 1399; (b) R. Tran-Son-Tay, B. B. Beaty, B. E. Coffey and R. M. Hochmuth, *10th Int. Cong. Rheo. Vol. 2*, P. H. T. Uhlherr (Ed.), 1988, Australian Society of Rheology, Sydney, p. 355; (c) R. Tran-Son-Tay, B. E. Coffey and R. M. Hochmuth, *J. Rheol.*, 1990, **34**, 169.
21. (a) M. Gahleitner and R. Sobczak, *Rheol. Acta*, 1987, **26**, 371; (b) K. Bernreitner, M. Gahleitner and R. Sobczak, *J. Non-Newtonian Fluid Mech.*, 1988, **30**, 73; (c) M. Gahleitner and R. Sobczak, *J. Phys. E Sci. Inst.*, 1988, **21**, 1074; (d) W. Hermann and R. Sobczak, *J. Appl. Polym. Sci.*, 1989, **37**, 2675; (e) M. Gahleitner and R. Sobczak, *Kunststoff German Plastics*, 1989, **79**, 67.

22. J. Happel and H. Brenner, *Low Reynolds Number Hydrodynamics*, Martinus Nijhoff, The Hague, 1983.
23. G. Dazhi and R. I. Tanner, *J. Non-Newtonian Fluid Mech.*, 1985, **17**, 1.
24. R. P. Chhabra, P. H. T. Uhlherr and D. V. Boger, *J. Non-Newtonian Fluid Mech.*, 1980, **6**, 187.
25. F. Sugeng and R. I. Tanner, *J. Non-Newtonian Fluid Mech.*, 1986, **20**, 281.
26. D. V. Boger, *J. Non-Newtonian Fluid Mech.*, 1977/78, **3**, 87.
27. N. H. Park and T. F. Irvine, *Rev. Sci. Instr.*, 1988, **59**, 2051.
28. W. Philippoff, C. D. Han, B. Barnett and M. J. Dulfano, *Biorheology*, 1970, **7**, 55.
29. J. Fisher and A. G. Fredrickson, *Mol. Crys. Liq. Crys.*, 1969, **8**, 267.
30. V. Seshadri and S. P. Sutera, *Trans. Soc. Rheol.*, 1970, **14**, 351.
31. R. J. Mannheimer, *J. Coll. Inter. Sci.*, 1972, **40**, 370.
32. T. Q. Jiang, A. C. Young and A. B. Metzner, *Rheol. Acta*, 1986, **25**, 397.
33. Y. Cohen and A. B. Metzner, *Rheol. Acta*, 1986, **25**, 28.
34. A. V. Ramamurthy, *J. Rheol.*, 1986, **30**, 337.
35. B. Y. Shin and I. J. Chung, *Polym. Bull.*, 1988, **20**, 399.
36. D. Y. C. Chan and R. G. Horn, *J. Chem. Phys.*, 1985, **83**, 5311.
37. (a) J. N. Israelachvili, *Pure Appl. Chem.*, 1988, **60**, 1473; (b) J. N. Israelachvili, S. J. Kott and L. J. Fetters, *J. Polym. Sci. Polym. Phys.*, 1989, **27**, 489.
38. J. van Alsten and S. Granick, *Phys. Rev. Lett.*, 1988, **61**, 2570; *Langmuir*, 1990, **6**, 687; *Macromolecules*, 1990, **23**, 4856.
39. (a) A. Tonck, J. M. Georges and J. L. Loubet, *J. Coll. Int. Sci.*, 1988, **126**, 150; (b) J. P. Montfort, A. Tonck, J. L. Loubet and J. M. Georges, *J. Polym. Sci. Polym. Phys.*, 1991, **29**, 677.
40. J. van Alsten, pers. comm., 1990.
41. J. van Alsten, S. Granick and J. N. Israelachvili, *J. Coll. Int. Sci.*, 1988, **125**, 739.
42. S. G. Hatzikiriakos and J. M. Dealy, *J. Rheol.*, 1991, **35**, 497.
43. M. E. Mackay and C. A. Cathey, *J. Rheol.*, 1991, **35**, 237.
44. H. M. Laun and J. Meissner, *Rheol. Acta*, 1980, **19**, 60.
45. J. M. Dealy and A. J. Giacomin, in *Rheological Measurements*, A. A. Collyer and D. W. Clegg (Eds), Elsevier Applied Science, London, 1988, p. 383.
46. W. J. Frith, J. Mewis and T. A. Strivens, *Powder Tech.*, 1987, **51**, 27; W. Stoks, H. Berghmans, P. Moldenaers and J. Mewis, *Brit. Polym. J.*, 1988, **20**, 361.
47. L. Dintenfass, *Biorheology*, 1963, **1**, 91; *Biorheology*, 1965, **2**, 221; *Biorheology*, 1969, **6**, 33.
48. D. O. Miles, *J. Appl. Phys.*, 1962, **33**, 1422; D. O. Miles and G. C. Knollman, *J. Appl. Phys.*, 1964, **35**, 2549; D. O. Miles, G. C. Knollman and A. S. Hamamoto, *Rev. Sci. Instr.*, 1965, **36**, 158.
49. H. C. Rorden and A. Grieco, *J. Appl. Phys.*, 1951, **22**, 842.
50. I. L. Hopkins, *Trans. Am. Soc. Mech. Eng.*, 1951, **73**, 195; *J. Appl. Phys.*, 1953, **24**, 1300.
51. J. P. Arnould, J. M. Zahm, G. Pottier, C. Duvivier and E. Puchelle, *Biorheology*, 1984, **Suppl. I**, 123.

Chapter 9

Rate- or Stress-Controlled Rheometry

WOLFGANG GLEIßLE

Institut für Mechanische Verfahrenstechnik und Mechanik, Universität Karlsruhe (TH), Karlsruhe, Germany

9.1. INTRODUCTION

Rheometry is progressively moving out of the scientific laboratory and entering into factories in manufacturing processes. In view of this new aspect, the production engineer and the manufacturer of viscometric instruments have to decide what kind of fundamental rheometric experiment would be the most successful, e.g. for process control, i.e. rate- or stress-controlled rheometry?

Both of the fundamentally different test techniques find use in applied rheometry. The selection of a particular viscometer for quality or process control, however, is usually based on its price and success in previous applications. As such, the nature of the principal experimental process is not taken into account.

The following notes describe and discuss the specific qualities of speed- and stress-controlled flow measurements and will try to give helpful hints for extracting as much information as possible from single-point on-line rheometry.

9.1.1. Examples for Applied Rheometry Today

A typical measuring instrument in the field of applied rheometry is the melt flow indexer, as schematically shown in Fig. 9.1.

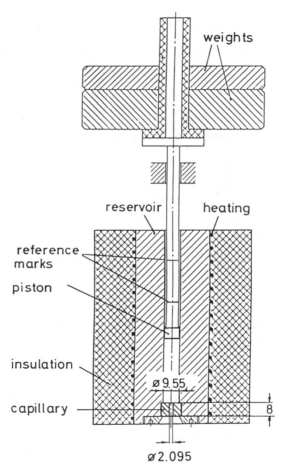

Fig. 9.1. Instrument for the investigation of the melt flow index as standardized in ISO 1133, DIN 53735 and ASTM D-1238.

The molten polymer in the reservoir is extruded through a capillary by the force of weights which load a piston. The measuring procedure and the dimensions of the reservoir, the piston and the capillary are standard-ized world wide. The mass or volume extruded through the capillary

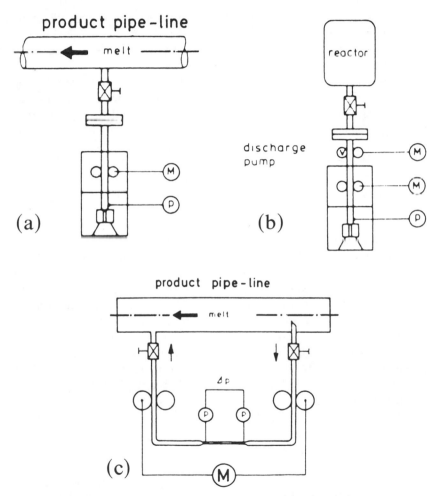

Fig. 9.2. Examples of 'on-line' adapted capillary rheometers . M = melt pump, P = pressure transducer. a, b: By-pass rheometers with lost sample stream; c: side-stream rheometer with two gear pumps incorporating sample recycling and pressure-independent viscosity measurement.

within ten minutes is the Melt Flow Index (MFI) or Melt Volume Rate (MVR). The MFI and MVR are the most common quality indexes which characterize the flow behavior of polymer melts with a single number (single-point measurement). The melt flow indexer is normally located in the laboratory and not directly at the production line; hence the MFI-value is determined 'off-line'.

To a first approximation, the MFI procedure is a stress-controlled flow experiment, controlled by the weights on the piston which create a constant pressure drop over the capillary. The amount of the extruded melt is measured to characterize the flow properties.

The standardized length L to diameter D ratio of the MVI capillary is, at about 3·8, extremely small. Polymer melts are, in most cases, liquids with a pronounced viscoelastic behavior. When forced through a capillary, significant entrance and outlet pressure losses arise. Because of the small L/D ratio and the viscoelastic effects, conclusions regarding the complex flow properties of the polymer melt from the MFI or MVR alone are problematical. A direct calculation of the viscosity from the MVR value leads in many cases to misinterpretations. (These problems are 'international' and will therefore be discussed in a separate chapter.) For all this, the melt flow-indexer is a successful quality-control instrument.

For immediate on-line rheometry, capillary rheometers have been proven in heavy-duty tasks. Typical examples of capillary rheometers which have been adapted to product lines and reactors are illustrated in Fig. 9.2. The output signals from such rheometers, directly mounted to reactors, can be used to control a production (e.g. a polymerization process), since there is no time lag between sampling and result output.

The volume flow rate through the capillary is controlled by a gear pump, which normally operates at constant speed of rotation. The pressure drop across the capillary is measured. Any pressure drop announces a change in the viscosity. This is a typical case of speed-controlled measurement.

9.2. THE PROBLEM

Polymer melts are pseudoplastic liquids with a pronounced viscosity drop with increasing shear rate. Typical examples for such decreasing viscosity functions are shown in Fig. 9.3. Their viscosity η depends strongly upon the shear rate $\dot{\gamma}$. This viscosity function $\eta(\dot{\gamma})$ normally cannot be characterized by a single-point measurement alone.

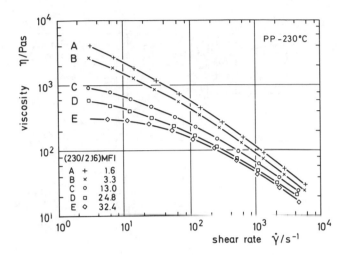

Fig. 9.3. Viscosity functions $\eta(\dot{\gamma})$ of five different polypropylenes A to E with the melt flow indexes 1·6–32·4.

A typical problem to be solved by the engineer who, for example, must control the production of polypropylenes, with the viscosity functions shown in Fig. 9.3, is to decide on the most suitable type of rheometer, the basic measuring procedure and the operating range with the best efficiency. If the production process must be regulated automatically by a viscometric method, he must determine that point of the viscosity function which he thinks represents the flow properties of that polymer the best.

Because of their robust construction, capillary viscometers are the most commonly applied type for on-line control and the choice of the most suitable viscometer would appear to be the simplest problem.

At first glance one can see that the differences between the single viscosity-functions in Fig. 9.3 are the greatest at low shear rates so that here, measurements of the viscosity should have the highest resolution.

9.3. RATE-CONTROLLED MEASUREMENTS

The special conditions of the measurement of the viscosity η at two different shear rates $\dot{\gamma}_1$ and $\dot{\gamma}_2$ will be commented on in the example of three PMMA grades, 5N, 7N and 8H. These viscosity functions are displayed in Fig. 9.4. The basic experiment is shear-rate–controlled, which implies a constant volume rate, i.e. $\dot{V}=$ constant, in a capillary or a con-

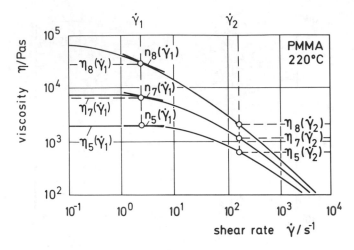

Fig. 9.4. Viscosity functions of different PMMAs: 5N, 7N and 8H.

stant angular speed in a rotational rheometer. At low shear rates, greater differences exist between the viscosity functions than at high shear rates. This also holds true for the relative as for the absolute values of the viscosities, given by eqns (9.1) and (9.2), where $\eta_8(\dot{\gamma}_1)$ represents the viscosity of the PMMA grade 8H at the shear rate $\dot{\gamma}_1$, and so on.

$$\frac{\eta_8(\dot{\gamma}_1)}{\eta_7(\dot{\gamma}_1)} > \frac{\eta_8(\dot{\gamma}_2)}{\eta_7(\dot{\gamma}_2)} \tag{9.1}$$

$$\eta_8(\dot{\gamma}_1) - \eta_1(\dot{\gamma}_1) > \eta_8(\dot{\gamma}_2) - \eta_7(\dot{\gamma}_2) \tag{9.2}$$

The viscosity functions differ to a greater extent at low shear rates than at high shear rates and the distinguishing resolution behaves equally. Unfortunately, however, the viscosities cannot be determined directly. In order to quantify η one must measure the shear stress or other physical quantities which are proportional to the shear stress in the form of a pressure drop in a capillary or the torque in a rotational rheometer.

In Fig. 9.5 one can observe the shear stress functions $\sigma(\dot{\gamma})$ belonging to the viscosity functions in Fig. 9.4 (log σ is plotted in dependence upon log $\dot{\gamma}$). According to the suitable viscosity functions in Fig. 9.4, the same distances between the shear stress functions can be found in Fig. 9.5. The relative resolution for the discrimination of the shear stresses is therefore greater at low than at high shear rates, as found for the viscosity in

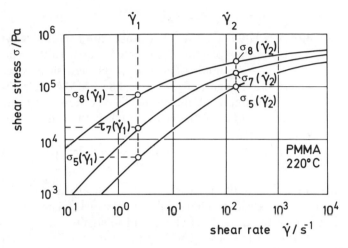

Fig. 9.5. Shear-stress functions of three PMMA grades, 5N, 7N and 8H.

eqn (9.3):

$$\sigma_8(\dot{\gamma}_1)/\sigma_7(\dot{\gamma}_1) > \sigma_8(\dot{\gamma}_2)/\sigma_7(\dot{\gamma}_2) \tag{9.3}$$

Since the absolute magnitudes of the shear stresses σ are small at low, and large at high shear rates, the absolute differences between the shear stresses are also smaller at low shear rates than at high shear rates (eqn (9.4)):

$$\sigma_8(\dot{\gamma}_1) - \sigma_7(\dot{\gamma}_1) < \sigma_8(\dot{\gamma}_2) - \sigma_7(\dot{\gamma}_2) \tag{9.4}$$

The result of this fact is that adverse conditions exist for the measurement of the pressure or torque at low shear rates because the measurement of, for example, low pressures are always problematic, especially at high temperatures and under rough industrial conditions.

Therefore, in spite of the worse relative resolution, the shear stress must normally be determined at elevated shear rates.

A fundamental problem concerning measurements of the flow properties of viscoelastic materials at constant shear rates, is the fact that such measurements are conducted under different rheological conditions. These different rheological conditions, in spite of constant shear rates, are manifested by different flow exponents n of the viscosity functions, as one may observe in Fig. 9.4 at $\dot{\gamma}_1$. The flow exponents n_i are the gradients of the viscosity curves at constant shear rate $\dot{\gamma}_i$.

At the shear rate $\dot{\gamma}_i = \dot{\gamma}_1$ one finds (Fig. 9.4):

Typ 5N: $n_5(\dot{\gamma}_1)$ is equal to zero. The fluid flows in the Newtonian range and the apparent shear rate is equal to the true shear rate: $\dot{\gamma}_{ap,1} = \dot{\gamma}_{w,1}$ and $\eta_5(\dot{\gamma}_1) = \eta_{5,0}$, the zero shear viscosity;

Typ 7N: $n_7(\dot{\gamma}_1)$ differs slightly from zero. The PMMA melt 7N flows at the beginning of the pseudoplastic flow range and the viscosity $\eta_7(\dot{\gamma}_1)$ is therefore smaller than the zero shear viscosity, $\eta_{7,0}$;

Typ 8H: $n_8(\dot{\gamma}_1)$ is far from zero and therefore the flow behavior is dominated by pseudoplastic effects.

These different flow conditions lead, for example, to non-similar flow profiles in the capillary flow (parabolic profile in the case of 5N and nearly plug-flow in the case of 8H at $\dot{\gamma}_1$) and to significant differences of the normal stresses in the flow. The Bagley-corrections in the capillary flow of 5N and 8H are totally different at the same shear rate $\dot{\gamma}_1$.

The result is: one cannot compare the flow phenomena of different fluids at a constant shear rate even if they are chemically very similar and differ only by their degree of polymerization. Rate-controlled measurements are very common in rheometry and are normally easy to conduct, but lead to difficulties in the interpretation of the results if just one single flow-curve point is measured.

9.4. STRESS-CONTROLLED MEASUREMENTS

Speed- or rate-controlled experiments (Mode I in Fig. 9.6) can be replaced by stress-controlled experiments (Mode II in Fig. 9.6). Stress control in a capillary experiment implies a constant pressure drop, and in rotational rheometers a constant torque.

As an example, a direct comparison of results from rate-controlled measurements with those from stress-controlled measurements is given in Fig. 9.6. Here, Mode I, is a vertical line and represents a constant shear rate, whilst Mode II is a straight line with a gradient of -1, which represents a constant shear stress. For the purpose of a quantitative comparison, the viscosities of the three PMMA melts at the constant shear rate $\dot{\gamma}_c = 10\ \text{s}^{-1}$ were compared with those at the constant shear stress $\sigma_c = 4 \cdot 5 \cdot 10^4\ \text{Pa}$ in Table 9.1.

From these results, measured in a central shear-stress or shear-rate region, one can conclude that the resolution should be higher for stress

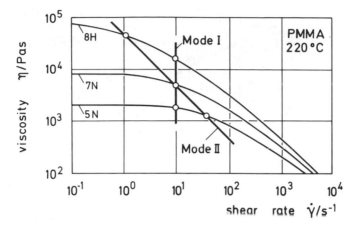

Fig. 9.6. Comparison between rate- (Mode I) and stress- (Mode II) controlled measurement.

controlled measurement than in the case of rate-controlled measurements. In this special case, the resolution is 4·5 times higher in Mode II than in Mode I.

If one compares the viscosities η_i and the flow exponents n_i of these three different PMMA melts measured under constant stress conditions at the shear stresses σ_1 and σ_2 (Fig. 9.7), one finds many 'similarities', and surprisingly simple relations emerge.

$$\eta_8(\sigma_1)/\eta_7(\sigma_1) = \eta_8(\sigma_2)/\eta_7(\sigma_2) \qquad (9.5)$$

Equation (9.5) demonstrates that the resolution for the discrimination of

TABLE 9.1

Comparison of the Viscosities of Three Different PMMAs Measured at Constant Shear Rate (Mode I) and at Constant Shear Stress (Mode II)

Mode I = rate-controlled $\dot{\gamma}_c = 10\ s^{-1} = constant$	*Mode II = stress-controlled* $\sigma_c = 4\cdot5\ Pa = constant$
$\eta_8(\dot{\gamma}_c) = 1\cdot4 \ . \ 10^4$ Pa s	$\eta_8(\sigma_c) = 4\cdot3 \ . \ 10^4$ Pa s
$\eta_7(\dot{\gamma}_c) = 5\cdot0 \ . \ 10^3$ Pa s	$\eta_7(\sigma_c) = 5\cdot0 \ . \ 10^3$ Pa s
$\eta_5(\dot{\gamma}_c) = 1\cdot8 \ . \ 10^3$ Pa s	$\eta_5(\sigma_c) = 1\cdot2 \ . \ 10^3$ Pa s
Resolution	
$\dfrac{\eta_8(\dot{\gamma}_c)}{\eta_5(\dot{\gamma}_c)} = 7\cdot9$	$\dfrac{\eta_8(\sigma_c)}{\eta_5(\sigma_c)} = 35\cdot6$

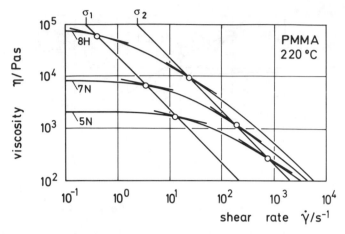

Fig. 9.7. Typical flow properties such as viscosity (circles) and flow exponent (tangents) corresponding to different constant shear stresses σ_1 and σ_2 of three PMMA melts 5N, 7N and 8H.

different viscosity functions is constant and independent of the magnitude of the shear stress σ. Equation (9.5) is, of course, also valid for all other combinations of viscosity ratios such as $\eta_8(\sigma_1)/\eta_5(\sigma_1)$, etc.

$$n_8(\sigma_1) = n_7(\sigma_1) = n_5(\sigma_1) \tag{9.6}$$

$$n_8(\sigma_2) = n_7(\sigma_2) = n_5(\sigma_2) \tag{9.7}$$

The flow exponent n is independent of the individual viscosity functions and is a constant for constant shear stresses (eqns (9.6) and (9.7)), provided constant shear-stress flow measurements are made under constant rheological conditions.

$$\frac{\dot{\gamma}_w}{\dot{\gamma}_{ap}} = \frac{1}{4}\left(3 + \frac{1}{n+1}\right) \tag{9.8}$$

As a consequence of the relations formulated in eqns (9.6) and (9.7), the Weißenberg–Rabinowitsch correction (eqn. (9.8)), which relates the true shear rate $\dot{\gamma}_w$ at the wall of a capillary to the apparent shear rate $\dot{\gamma}_{ap}$ calculated from the Hagen–Poiseuille equation, is constant at a constant shear stress since eqn. (9.8) depends upon the flow exponent n alone. Therefore one will find similar radial velocity profiles $v(r)$ in capillaries at constant wall shear stress.

If eqn (9.5) holds, then stress-controlled viscosity measurements can be made at any shear stress with the same (maximal) relative sensitivity. In practical terms this means that viscosity control can be made within this range with the most advantageous measuring conditions of the available rheometer. Equation (9.5) is also valid at extremely low shear stresses where the viscosity is independent of the shear rate. Therefore the change of the flow rate $\dot{V}$ through a capillary at a constant pressure loss Δp (or the change of the angular velocity ω in a rotational rheometer at constant torque M) is inversely proportional to the change of the zero-shear viscosity η_0. This fact is formulated in eqn (9.9)

$$\frac{\dot{V}_I(\Delta p)}{\dot{V}_J(\Delta p)} = \frac{\omega_I(M)}{\omega_J(M)} = \frac{\dot{\gamma}_I(\sigma)}{\dot{\gamma}_J(\sigma)} = \frac{\eta_{0,J}}{\eta_{0,I}} \tag{9.9}$$

where the indices I and J indicate two different fluids.

9.5. VISCOUS AND VISCOELASTIC SIMILARITY

The background for these simple equations which relate the viscosities and the flow exponents is based on a physical property which one can name 'viscous similarity'.

Fluids are viscous similar if their viscosity functions have the same shape, especially if their individual flow functions can be transformed by simple transformation conditions into one single master curve.

Viscous similarity can be proved, for example, by plotting the reduced viscosity $\eta_r = \eta/\eta_0$ as a function of the 'reduced' shear rate $\eta_0 \cdot \dot{\gamma}$ which has the dimension of a shear stress, designated here as 'Newtonian' shear stress σ_N. This formulation of an invariant viscosity function was proposed by Vinogradov and Malkin in the 1960s.[1]

As can be seen in Fig. 9.8, the viscosity functions of the three PMMA melts are represented by a single master-curve. These PMMA melts are therefore 'viscous similar' fluids.

Each point on this reduced viscosity function represents a constant shear stress σ_c, which is constant for all individual viscosity functions, as specified in eqn. (9.10)

$$\sigma_c = \eta_r \sigma_N = (\eta/\eta_0) \cdot \eta_0 \cdot \dot{\gamma} = \eta\dot{\gamma} \tag{9.10}$$

The class of thermo-rheologic simple fluids is included in the viscous similarity.

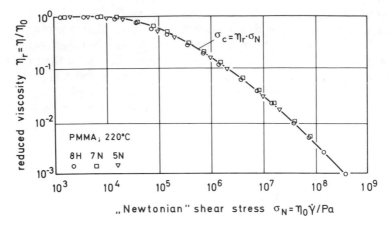

Fig. 9.8. Reduced viscosity function of different PMMA melts.

Polymer melts are pseudoplastic liquids which additionally reveal pronounced elastic effects. They are viscoelastic. The elastic properties of a liquid are described by the first normal stress coefficient ψ_1. If the viscosity function $\eta(\dot{\gamma})$ and the function of the first normal stress coefficient $\psi_1(\dot{\gamma})$ are coupled by such simple relations as proposed from Wagner,[2] (eqn (9.11)) or Gleißle,[3,4] (eqn (9.12)), then the shear stress σ and the first normal stress difference N_1 are uniquely related. 'Viscous-similar' fluids are then 'viscoelastic-similar' fluids.

Wagner:
$$\psi_1(\dot{\gamma}) = -\frac{1}{n_D} \cdot \frac{d\eta(\dot{\gamma})}{d\dot{\gamma}}$$
(9.11)

Gleißle:
$$\psi_1(\dot{\gamma}) = 2 \int_{\eta_\infty}^{\eta(\dot{\gamma}/k)} \frac{d\eta}{\dot{\gamma}}$$
(9.12)

The viscoelastic similarity of the PMMA melts can be observed in Fig. 9.9, where σ and N_1 are plotted as a function the 'Newtonian' shear stress (= reduced shear rate).

The function of the first normal stress coefficient $\psi_1(\dot{\gamma})$ is defined in eqn (9.13) with the help of the first normal stress difference $N_1(\dot{\gamma})$.

$$\psi_1(\dot{\gamma}) = N_1(\dot{\gamma})/\dot{\gamma}^2$$
(9.13)

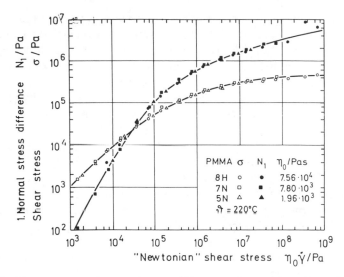

Fig. 9.9. Shear stress and normal stress difference of PMMA melts as functions of the 'Newtonian' shear stress.

For a class of 'viscoelastic-similar' liquids, the shear stress and the first normal stress difference are related by a single function (see eqn (9.14)).

$$N_1 = f(\sigma) \tag{9.14}$$

9.6. VISCOELASTIC SIMILARITY AND BAGLEY-CORRECTION

In the 1950s, the rheologists had to learn that the capillary experiment for the investigation into the viscosity of polymer melts is not as trivial as presumed.

As an introduction in Ref. 5, Philippoff wrote: 'The use of the flow through a cylindrical tube for the measurement of the viscosity of liquids was introduced by Poiseuille in 1840 and has been used ever since. However, recently we have come across phenomena that show how careful one must be when utilizing this well-known experiment for the determination of the rheological properties of viscoelastic liquids. This led to a re-evaluation of some of the fundamental principles of rheology that have been generally accepted during the past 30 years.'

The phenomenon which the rheologists encountered was that a linear relationship could not be measured between the real capillary length and the total pressure drop for viscoelastic liquids even at very low Reynolds numbers. Unusually large entrance and outlet pressure losses occur when forcing viscoelastic liquids through a capillary.

In 1957 Bagley proposed to take these effects into account by assuming an effective capillary length $L + nR$, i.e. greater than the actual capillary length. He demonstrated with an LDPE melt that the total pressure drop Δp measured with different capillary lengths at constant wall shear rate $\dot{\gamma}_w$ yielded a linear relationship if plotted over the reduced capillary length L/R (Fig. 9.10). By extrapolation of this straight line to $\Delta p = 0$, the value n becomes apparent. The plot of Δp as a function of L/R is known as the Bagley-function, and the method of correcting the capillary measurements in order to obtain the true wall shear stress σ_w is hence the Bagley-correction.

A slightly modified method is to take the viscoelastic pressure loss into account as an additional pressure loss Δp_E according Fig. 9.11. The applied pressure Δp comprises two terms: the viscous pressure loss Δp_V,

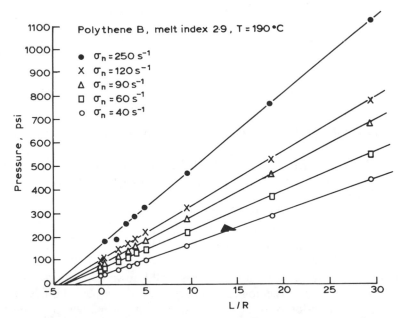

Fig. 9.10. Reproduction of Fig. 4 of Bagley's original paper.[6]

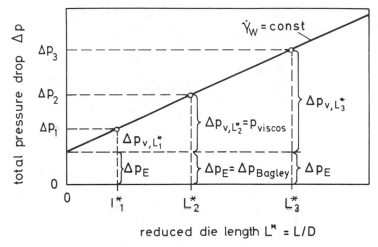

reduced die length $L^* = L/D$

Fig. 9.11. Scheme of the pressure drop of a viscoelastic liquid along a capillary (Bagley plot).

generated by the viscosity of the fluid, and which is proportional to the length of the die, and the extra pressure or elastic pressure loss Δp_E, generated by the elastic properties of the liquid, which is a constant at constant wall shear rate (eqn (9.15)). The acting wall shear stress must be calculated from the viscous pressure loss Δp_V only (see eqn (9.16)), which is the total pressure loss Δp from which the elastic pressure Δp_E is subtracted.

$$\Delta p(L) = \Delta p_V(L) + \Delta p_E \tag{9.15}$$

$$\sigma_w = \frac{\Delta p_V(L)}{2L/R} = \frac{\Delta p(L) - \Delta p_E}{2L/R} \tag{9.16}$$

The total pressure drop as function of the die length, as measured for an LDPE melt at 190°C is shown in Fig. 9.12, where a strongly increasing elastic pressure loss with increasing shear rate parameter $\dot\gamma$ can be observed at $L = 0$. For the highest shear rate and the die length of 8 mm, one may observe that one half of the total pressure drop is consumed by the elastic pressure loss (Bagley-correction).

Considering these problems, the question is: can experiments with different liquids be conducted at a constant shear stress σ_w by simply holding the total pressure drop constant over a capillary.

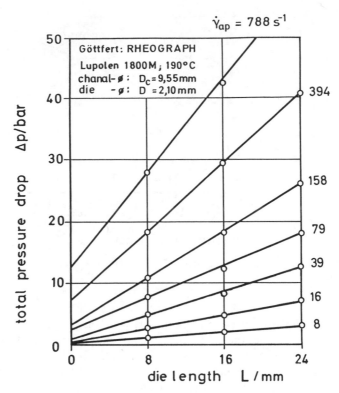

Fig. 9.12. Total pressure drop as function of capillary length L for an LDPE melt at different shear rates $\dot{\gamma}$.

In Fig. 9.13 the shear stress σ, the normal stress difference N_1 and the Bagley-pressure of LDPE 1800 S are plotted in dependence on the shear rate. This plot demonstrates that the experimentally determined Bagley-pressure Δp_E is proportional to the first normal stress difference, and that for this LDPE melt the absolute value of the Bagley-pressure is 5 times higher than the first normal stress difference.[7] Similar proportionalities between Δp_E and N_1 were found for many other viscoelastic liquids (eqn (9.17)), though with different factors q.

$$\Delta p_E = q \cdot N_1 \qquad (9.17)$$

If the Bagley-pressure and the first normal stress difference possess the mutual proportionality formulated in eqn (9.17), then for viscoelastic liquids the shear stress and the Bagley-correction are also directly coupled

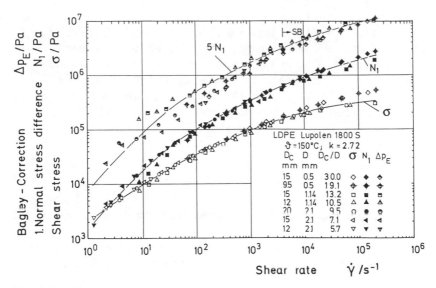

Fig. 9.13. Shear stress, first normal stress difference and Bagley-correction pressure as functions of $\dot{\gamma}$. D_c = diameter of the reservoir of the capillary rheometer, D = diameter of the die, D_c/D = entrance ratio.

(eqn (9.14) and eqn (9.17)). The elastic pressure losses Δp_E and the shear stress σ are related to each other by eqn (9.18) for viscoelastic-similar fluids, independent of the individual viscosity functions.

$$\Delta p_E = f(\sigma) \qquad (9.18)$$

In a measuring device with a capillary of length L and radius R, used, for example, to control a process at constant pressure drop, the Bagley-pressure is independent of the actual volume flow rate and is only dependent upon the shear stress, or the respective pressure drop $p(L)$, as demonstrated in eqn (9.15): Implementing eqns (9.16) and (9.17), one obtains:

$$\Delta p(L) = \Delta p_V(L) + \Delta p_E = \frac{2L}{R} \cdot \sigma + q \cdot N_1 \qquad (9.19)$$

and

$$\Delta p(L) = \frac{2L}{R} \cdot \eta \cdot \dot{\gamma} + q \cdot \psi_1 \cdot \dot{\gamma}^2 \qquad (9.20)$$

Under pressure-controlled conditions, the total pressure drop $p(L)$ in a capillary may be split into two factors: the viscous pressure drop $\Delta p_V(L)$ and the elastic pressure drop Δp_E. At constant capillary length L, the ratio between $\Delta p_V(L)$ and Δp_E is independent of the shear rate and is just a function of σ. When forced through a capillary rheometer, operated at constant pressure drop, viscoelastic-similar liquids with varying viscosities η will reveal changes in the volume rate which are directly proportional to the variation of the viscosity, since the elastic factor of the total pressure drop remains constant (see eqn (9.21)):

$$\Delta p(L) = \frac{2L}{R} \cdot \eta \cdot \dot{\gamma} + \Delta p_E \qquad (9.21)$$

At a constant pressure drop, the liquid elasticity does not violate the measuring conditions for the viscosity. Also, in a melt flow indexer, the change of MFI is inversely proportional to the change in the viscosity, if viscoelastic-similar liquids are tested under the same loading.

9.7. EXPERIMENTS

The equations relating the viscosities $\eta_i(\sigma_i)$ and flow exponents $n(\sigma_i)$ of viscoelastic-similar fluids measured at constant shear stress, can be checked by the flow functions of a variety of different polypropylene types produced on the same reaction line. The melt flow indices of these different PP grades A–E vary from MFI = 1·6 to MFI = 32·4 g/10 min. Their viscosity functions were investigated with a capillary rheometer (RHEO-GRAPH 2002, Göttfert, Rheologische Prüfgeräte, W-6967 Büchen, Germany) over an extended range of shear rates (Fig. 9.14) and their flow properties were compared at different constant shear stresses σ_1 and σ_2.

In addition to this comparison between the flow curves of the individual PP-types measured with a fundamental (laboratory) rheometer, the same fluids were tested in a new type of side-stream capillary rheometer, working at constant pressure loss directly flanged to an extruder. Under these operating conditions, the on-line rheometer pump's speed should be proportional to the MFI values of the individual PP types A–E if all presumptions formulated in eqns (9.5)–(9.21) are valid.

If the viscosity functions of the polypropylenes in Fig. 9.15 are viscous-similar, then they should be representable by a single master curve. Because the zero shear viscosities of these polymers are unknown,

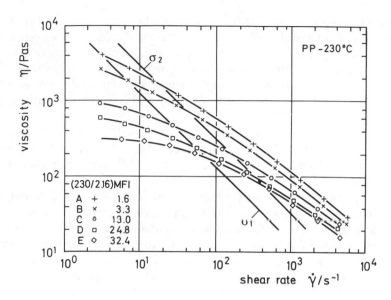

Fig. 9.14. Viscosity function of different polypropylenes.

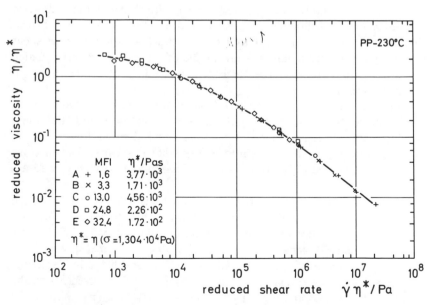

Fig. 9.15. 'Reduced' viscosity functions of polypropylene.

however, a reduction in that form proposed by Vinogradov and Malkin (Fig. 9.8) is not possible. To overcome these difficulties and to obtain a single viscosity function to prove viscous similarity, the viscosity $\eta^*(\sigma_1)$ at a constant shear stress σ_1 was used (which is, of course, within the measured range of all the viscosity functions) to reduce the viscosity as well as the shear rate. The result of this special form of 'normalization' is presented in Fig. 9.15.

The reduction of the viscosity functions of the PP types A–E with the individual viscosities $\eta^*(\sigma_1)$ at the shear stress $\sigma_1 = 1 \cdot 304 \cdot 10^4$ Pa indeed leads to a uniform viscosity function for all PP melts as included in Fig. 9.15.

Unlike that of the reduced viscosity function $\eta_r = \eta/\eta_0$ according to the procedure of Vinogradov and Malkin, by this kind of reduction the maximum value of the reduced viscosity function exceeds unity. However, this special kind of 'normalization' can lead to an approximative determination of the individual zero shear viscosities. In the special case of the PPs, the individual values of η_0 should be approximately $2 \cdot 5$ times higher than the individual value of $\eta^*(\sigma_1 = 1 \cdot 3 \cdot 10^4$ Pa$)$ ($2 \cdot 5$ is the asymptotic value for all of these reduced viscosity functions at low shear rates).

In Fig. 9.16 the shear stresses of the polypropylene melts A to E are shown as functions of the apparent shear rate $\dot{\gamma}_{ap}$ together with two straight lines of constant shear stress σ_1 and σ_2. The magnitudes of these shear stresses σ_1 and σ_2 are equal to two-thirds of the apparent shear stresses generated nominally within the die of a melt flow indexer loaded with $M_0 = 2 \cdot 16$ or $5 \cdot 0$ kg, respectively (M_0 is the mass of the death weight as can be seen in Fig. 9.1).

To check the concept, that rheometric control measurements can be conducted at any shear stress with the same sensitivity and that these measurements are representative for a complete individual flow function, the shear rates $\dot{\gamma}_{ap}$ found at the intersections of the shear stress functions A to E with the lines of constant shear stress σ_1 and σ_2 in Fig. 9.16 are listed in Table 9.2.

The melt flow index MFI is the result of a measuring procedure which is standardized over the globe and which agrees to a first approximation with a stress-controlled flow test. To obtain a reference value of the flow behavior of the PP-melts A–E, their MFI-values were measured with a melt flow indexer at $M_0 = 2 \cdot 16$ kg and $T = 230°C$ and listed in Table 9.2. If all assumptions formulated in eqns (9.5)–(9.21) are correct, then the ratios of the corresponding shear rates $\dot{\gamma}_{ap,I}(\sigma_1)/\dot{\gamma}_{ap,J}(\sigma_1)$ should be independent of the applied shear stress. To facilitate a direct comparison,

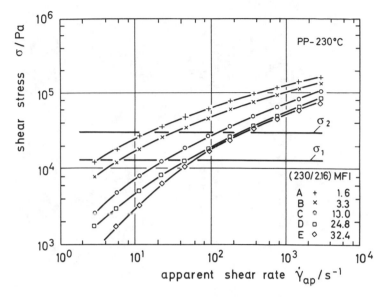

Fig. 9.16. Shear-stress functions of different polypropylenes.

the apparent shear rate $\dot{\gamma}_{\mathrm{ap,D}}(\sigma_i)$ of PP grade D was related with its own melt flow index $\mathrm{MFI_D}$. The corresponding melt flow indices $\mathrm{MFI}_{\mathrm{cal},X}$ of the other PP-melts were then calculated from eqn (9.22) (reference values in Table 9.2 are underlined).

$$\mathrm{MFI}_{\mathrm{cal},X} = \mathrm{MFI_D} \cdot \frac{\dot{\gamma}_{\mathrm{ap},X}}{\dot{\gamma}_{\mathrm{ap,D}}} \tag{9.22}$$

At the reference shear stress σ_1 one finds complete agreement between the MFIs measured with the standard melt flow indexer and the $\mathrm{MFI}_{\mathrm{cal}}$ calculated with eqn (9.22) from the apparent shear rates. Consequently, the standard MFI values can be interrelated by the shear stress function alone, even when measured in the presence of viscoelastic effects (e.g. Bagley-pressure). Conversely, it is possible to relate one shear-stress function with the other using the ratio of the melt flow indexes or the shear rates as a shift factor. The shear rates (or volume rates in a capillary) measured at constant shear stress (or pressure drop) are representative of a complete shear-stress function. This result additionally demonstrates that the MFI-measuring procedure is actually stress-controlled.

TABLE 9.2

Comparison of Shear Rates Measured with a Capillary Rheometer, MFI Values and On-Line Measurements of MFI Values of Different PP Types at Two Different Shear Stresses σ_1 and σ_2 Together with the Flow Exponents, m

Reference stress		$\sigma_1 = 1\cdot3 \, . \, 10^4 \, Pa$			$\sigma_2 = 3\cdot0 \, . \, 10^4 \, Pa$			$\sigma_1 = 1\cdot3 \, . \, 10^4 \, Pa$
Fluid (PP)	*MFI* (230/2·16) *melt flow indexer* (g/10 min)	$\dot{\gamma}_{ap}$ (s^{-1})	MFI_{cal} (g/10 min)	*Flow exponent* $m = \dfrac{d \log \sigma}{d \log \dot{\gamma}}$	$\dot{\gamma}_{ap}$ (s^{-1})	MFI_{cal} (g/10 min)	*Flow exponent* $m = \dfrac{d \log \sigma}{d \log \dot{\gamma}}$	*MFI–RTR On-line rheometer* L/D=25 (g/10 min)
A	1·6	3·0	1·64	0·61	13·8	1·55	0·49	1·51–1·65
B	3·3	6·1	3·32	0·60	29·6	3·31	0·47	3·14–3·40
C	13·0	23·1	12·6	0·61	108·0	12·1	0·48	
D	24·8	45·5	24·8	0·59	222·0	24·8	0·48	24·0–25·5
E	32·4	60·0	32·7	0·62	253·0	28·1	0·54	32·1–32·6

The flow exponents $m = n + 1$, listed in Table 9.2, were calculated from the shear-stress functions. As a consequence of the viscous similarity of all PP-grades at σ_1 their flow exponents $m(\sigma_1)$ agree with each other within the accuracy of the measurement, as predicted from eqn (9.6).

A slightly different situation may be found at σ_2. A good agreement between the measured MFI and the calculated MFI_{cal} was only existent for the PP grades A–D. Using eqn (9.22), $\dot{\gamma}_{ap,E}(\sigma_2)$ and MFI_D as reference index, a melt flow index $\text{MFI}_{cal,E}$ was calculated which is too small in comparison to its standard MFI value. Also, the flow exponent n differs significantly from the others at the same shear stress σ_2. A glance at Fig. 9.16 confirms that the shear-stress function of PP_E appears closer to its neighboring shear-stress functions PP_D at σ_2 then at σ_1. The viscoelastic similarity of PP grade E to the other PP grades A–D is thus not quite exact.

To test viscoelastic similarity, flow measurements should at least be made at two different shear stresses. In the case of similar fluids, the shear rates change proportionally to each other (eqn (9.23)) independent of the acting shear stress.

$$\frac{\dot{\gamma}_I(\sigma_1)}{\dot{\gamma}_J(\sigma_1)} = \frac{\dot{\gamma}_I(\sigma_2)}{\dot{\gamma}_J(\sigma_2)} \qquad (9.23)$$

Equation (9.23) is valid for both true as well as for apparent shear rates.

To prove the results obtained from viscometric measurements made off-line in the laboratory with a melt flow indexer and with a capillary rheometer (RHEOGRAPH 2002) the same PP grades were melted in an extruder and tested over a period of some weeks on-line with a capillary rheometer (RTR; Göttfert Rheologische Prüfwaschinen, W-6967 Büchen, Germany, Fig. 9.17), working at constant pressure drop. The type of melt forced through the extruder was changed in periods of some hours. The result, which is shown as the last column of Table 9.2, demonstrates the very narrow range of time-dependent variation of the MFI–RTR values, calculated from the volume rate, and the good agreement with the original MFIs.

9.8. CONCLUSIONS

For the above-mentioned reasons, stress-controlled measurements possess significant advantages in comparison to rate-controlled measurements, especially when applied to process and quality control.

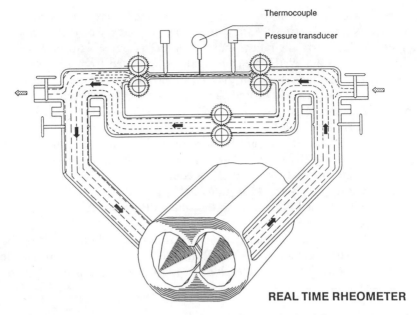

Fig. 9.17. Real Time Rheometer RTR (Gleißle, Göttfert) for stress or rate-controlled on-line measurements.

Rate-controlled measurements
Advantages

● Unproblematical rheometer control coupled with low costs.
● Common measuring procedure and data processing.
● Robust instrumentation, tested under heavy-duty conditions.

Disadvantages

● Working point-dependent sensitivity.
● Measurements under different rheological conditions (velocity profiles, normal stresses) even for viscoelastic-similar fluids.
● Single-point measurements are not representative for the flow function.

Stress-controlled measurements
Advantages

● High sensitivity.

- Sensitivity independent of the working point for viscous and visco-elastic-similar fluids.
- Measurements under constant rheological conditions (velocity profile, elastic effects) for viscoelastic similarity.
- Single-point measurements representative for the whole flow function.
- Shear-rate variation proportional to the zero-shear viscosity.

Disadvantages

- More expensive rheometer control system.

REFERENCES

1. G. V. Vinogradov and A. Ya. Malkin, *J. Polym. Sci. Pt. A-2*, 1966, **4**, 135.
2. M. H. Wagner, *Rheol. Acta*, 1979, **18**, 33.
3. W. Gleißle, Two simple time–shear rate relations combining viscosity and first normal stress coefficient, in *Rheology*, G. Astarita, G. Marrucci and L. Nicolais (Eds.), (Proc. VIIIth Int. Congr. on Rheology, Naples, 1980), Plenum Press, New York and London, 1980, p. 457.
4. W. Gleißle, *Rheol. Acta*, 1982, **21**, 484.
5. W. Philippoff and F. H. Gaskins, *Trans. Soc. Rheol.* 1958, **II**, 263.
6. E. B. Bagley, *J. Appl. Phys.*, 1957, **28**(5), 624.
7. W. Gleißle, First normal stress difference and Bagley-correction, in *Proc. Xth Int. Congr. on Rheology*, Vol. 1, P. H. T. Uhlherr (Ed.), Sydney, 1988, Australian Society of Rheology, p. 350.

Chapter 10

Rheometry for Process Control

J. M. DEALY AND T. O. BROADHEAD

*Department of Chemical Engineering, McGill University,
Montreal, Quebec, Canada*

10.1. INTRODUCTION

10.1.1. Applications of Rheometers in Manufacturing Operations

Rheometers, mostly viscometers, have been used for many years as sensors to monitor the characteristics of a fluid while it is being processed. While simpler probes, such as temperature sensors and pressure transducers, are widely used to measure process conditions, except when thermodynamic equilibria are involved, these variables provide no information about the composition or consistency of the material being processed. Obviously, the availability of such information, without the need for drawing samples and taking them to a laboratory, is very useful for quality control. However, an application that has a much greater potential for improving product quality and process efficiency is process control.

The first law of process control is that the quality of the control action can be no better than the quality of the data available from the process sensors. Thus, no matter how sophisticated the controller, its ability to maintain a uniform product quality is limited by the information available from the process sensors. In the processing of a great variety of materials, from plastics to foodstuffs, from paint to mineral slurries, rheological properties are often directly related to material characteristics of primary interest such as composition, molecular weight, degree of reaction or degree of mixing. Sometimes, as in the case of fuel oils and certain plastics molding compounds, the viscosity itself is the characteristic of primary interest. In other cases, it is necessary to establish the relationship between the rheological property measured and the characteristic of primary interest, in order to make proper use of the output signal of a rheological sensor.

10.1.2. Basic Elements of a Rheological Measurement

Rheological properties reflect the relationship between the deformation of a material and the force associated with that deformation. For example, the force required to stretch a rubber band is governed by the stiffness or 'elastic modulus' of the rubber, which is a rheological property. In the case of a motor oil, the shear force resulting from the shearing of the oil in the narrow gap between a cylinder wall and a piston is governed by the viscosity of the oil; viscosity is another rheological property. A 'thicker' oil, having a higher viscosity, will require more force to shear it. Many materials of commercial importance have rheological properties that are more complicated than a simple rubber or a Newtonian liquid. Examples are molten plastics, which are viscoelastic, and concentrated suspensions, which may behave like solids up to a certain stress and then flow like fluids.

Clearly, to learn something about the rheological properties of a material, it is necessary to deform the material in some controlled way and to measure the force required to generate that deformation. Alternatively, the deformation required to generate a certain force could be measured. In either case, this requires a more elaborate instrument than that required to measure a process condition such as temperature or pressure. In particular, the sensors used to measure these latter quantities are 'passive' in the sense that the material being processed is not affected by their presence. However, to make a rheological measurement, it is necessary for the sensor to interact with the material in an active way, in particular, by deforming it. There can be no rheological measurement without deformation. For this reason, the design of rheometers to be used as process sensors is a challenging field.

10.1.3. On-Line *versus* In-Line Installation
It will be useful to distinguish between 'in-line', 'on-line', and immersion units and to contrast these with 'off-line' measurements. An in-line rheometer is installed directly in the process line. The advantage is that sample renewal is generated by the main process flow and the signal delay is quite small. The disadvantage is that it is not possible to control the condition of the fluid prior to making the rheological measurement. In other words, the temperature, pressure and flow rate in the rheometer are governed by the process and are subject to variation. The principal complication here is that since rheological properties are strongly dependent upon temperature, any variation in temperature must be taken into account in the interpretation of the sensor output signal. Other special problems associated with in-line measurements are that the sensor must not interfere with the process and the process flow must not interfere with the operation of the rheometer.

Few types of rheometer are suitable for true in-line operation, and most commercial units are of the on-line type. This means that a side stream of the fluid is drawn from the main process flow, usually by means of a gear pump. This sample stream then passes to the rheometer. After the measurement has been made, the fluid can be returned to the main process line. For very viscous fluids, for example molten plastics, a second gear pump may be required to accomplish this. In order to avoid this additional complexity, the fluid leaving the rheometer is often allowed to exit the process as a waste stream.

The use of a side stream makes it possible to condition the sample, i.e. to control its temperature, pressure and flow rate. This obviates the need for temperature compensation. However, the time required for the sample to flow from the main flow line to the point where the rheological measurement is actually made causes a major delay in obtaining information about the state of the fluid at the sampling point. Except in rather slow processes, this long delay makes on-line rheometers unsuitable for use as sensors for process control.

Another type of installation is an immersion unit, designed for use in a process vessel. Finally, in-line, on-line and immersion devices are to be contrasted with 'off-line' measurements, i.e. those made in a laboratory on a discrete sample taken manually using a sampling valve. Whereas the turnaround time for results of off-line testing can be as much as an hour or more, even rather slow process rheometers can produce a signal in a matter of minutes, and in-line units can respond in seconds.

10.2. RHEOLOGICAL PROPERTIES OF FLUIDS

Before looking in detail at the problems involved in the design of a process rheometer, it will be useful to survey the rheological behavior exhibited by the fluids encountered in processing operations.

10.2.1. Shear Strain, Shear Rate and Shear Stress

Most rheometers, whether for use in the laboratory or as process sensors, make use of a shear deformation; the quantity measured is a force, pressure drop or torque that is directly related to the shear stress. For this reason, it will be useful to discuss material behavior in terms of this type of deformation. For the simplest type of rheological behavior, i.e. Newtonian fluid behavior, a single such measurement gives us a complete rheological characterization of the material. While this is not the case for more complex materials, such as multiphase systems and polymeric liquids, it will still be useful to base this introductory discussion on the behavior exhibited by various materials in a shear deformation.

The simplest type of shear deformation is 'simple shear', which is the deformation generated when a material is placed between two parallel flat plates and one of the two plates is then displaced linearly, keeping the gap between the plates constant. This arrangement is sketched in Fig. 10.1. If this gap is h and the linear displacement of the moving plate is Δx, then a measure of the deformation so generated is the 'shear strain', γ, given by

$$\gamma = \Delta X / h \tag{10.1}$$

If the plate moves at a constant speed, V, then the 'shear rate', $\dot{\gamma}$ is

$$\dot{\gamma} = V/h \tag{10.2}$$

The quantity measured is the shear stress, σ, defined as the force required to move the moving plate, divided by the area of the plate wetted by the material being deformed:

$$\sigma = F/A \tag{10.3}$$

Such an arrangement can be used to measure rheological properties by displacing the moving plate in some prescribed way and measuring the resulting shear stress. The rheological behavior can then be presented by giving the relationship between the stress and the shear strain or the shear rate. This relationship may be very simple, as for a Newtonian fluid, or it may be rather complex, in the case of bread dough or molten plastic.

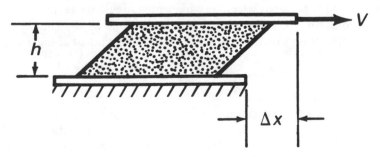

Fig. 10.1. Generation of simple shear. Fluid is contained between a lower, stationary plate and an upper, moving plate. The gap between the two plates is h, the displacement of the moving plate is ΔX, and the velocity of the moving plate is V.

In any case, the determination of the relationship between stress and deformation is the central element of any rheological measurement.

For example, if the material of interest is a purely elastic rubber, and if the strain is sufficiently small, then when the stresses required to generate shear strains of various sizes are measured, it will be found that the shear stress is proportional to the shear strain. Thus, the rheological behavior can be described by the following equation:

$$\sigma = G\gamma \qquad (10.4)$$

The constant G is the shear modulus, which in this example depends upon the composition and crosslink density of the rubber, but not on γ.

10.2.2. Newtonian Fluids

Materials consisting of a single liquid phase and containing only low molecular weight, mutually soluble components, are always Newtonian. Examples are water, gasoline, glycerine, ethylene glycol (anti-freeze) and ethanol. Miscible mixtures of Newtonian fluids are also Newtonian in behavior. For a Newtonian fluid, the shear stress is always proportional to the shear rate. This can be expressed quantitatively by means of the following equation:

$$\sigma = \eta\dot{\gamma} \qquad (10.5)$$

where η is the viscosity of the fluid. For a Newtonian fluid the viscosity depends upon composition and temperature but not on the shear rate, $\dot{\gamma}$. The viscosity also depends upon pressure, but this is a weak effect and need only be considered when very high pressures are encountered.

To summarize, the behavior of a Newtonian fluid to any kind of deformation depends only on its viscosity, which is independent of the deformation rate.

10.2.3. Non-Newtonian Fluids

Many materials processed commercially are multiphase fluids. Examples are fermentation broths, mineral slurries, paints and foodstuffs. Another important category of commercially important material is polymeric liquids, either polymer solutions or molten plastics. All of these materials are non-Newtonian. The simplest manifestation of this is that in steady simple shear, the viscosity varies with the shear rate. Thus, we have

$$\sigma/\dot{\gamma} = \eta(\dot{\gamma}) \qquad (10.6)$$

The most common type of behavior is when the viscosity decreases as the shear rate increases, and such a material is said to be 'shear thinning' (an older terminology was 'pseudoplastic'). Some concentrated suspensions can exhibit the opposite type of behavior with η increasing with $\dot\gamma$, which is called 'shear thickening' (formerly called 'dilatancy').

A simple, empirical equation for describing the dependence of viscosity on shear rate over a certain range of shear rates is the 'power-law' viscosity model shown below:

$$\eta = k\dot\gamma^{n-1} \tag{10.7}$$

where n is the power-law index. Note that when $n=1$, Newtonian behavior is indicated, while $n<1$ implies shear-thinning behavior. It is also important to note that this is an empirical equation and that it describes the viscosity of certain fluids only over a certain range of shear rates.

Figure 10.2 shows the viscosity as a function of shear rate on a log–log plot for a shear-thinning fluid. At moderate-to-high shear rates, the fluid viscosity follows power-law behavior while at very low shear rates the fluid viscosity is constant. At very high shear rates, there may be a second region in which the viscosity is constant.

A concentrated suspension can also have a viscosity that varies with the length of time it has been sheared. When the viscosity decreases with time, the material is said to be 'thixotropic'. This structural time-dependency is a major challenge for process designers. Moreover, it complicates the interpretation of rheological data in terms of basic material characteristics.

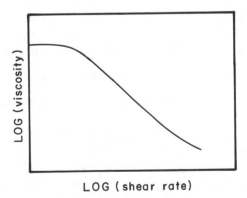

LOG (shear rate)

Fig. 10.2. Log–log plot of viscosity as a function of shear rate for a typical shear-thinning fluid. In the intermediate range of shear rates, the fluid exhibits power-law behavior.

10.2.4. Viscoelastic Fluids

Polymeric liquids exhibit a combination of elasticity and viscous flow and are therefore said to be 'viscoelastic'. A consequence of this is that their rheological properties are time-dependent. However, this viscoelastic time-dependency is different from the structural time-dependency mentioned above in the discussion of concentrated suspensions. In particular, the stress does not fall to zero immediately after deformation ceases, as it does for an inelastic liquid. Furthermore, some of the energy that goes into deforming the material is stored elastically, and this results in recoil when deforming stresses are suddenly removed.

The viscoelastic properties of a polymeric fluid, in particular the storage and loss moduli defined below, can be very useful as measures of molecular weight distribution or as measures of the extent of dispersion of a particulate filler.[1]

Fluid viscoelasticity is usually described in terms of the response of the fluid to a sinusoidal shearing in which the shear strain is given by

$$\gamma(t) = \gamma_0 \sin(\omega t) \tag{10.8}$$

where ω is the frequency of the oscillatory strain (deformation) and γ_0 is the strain amplitude. If the strain amplitude is sufficiently small, the shear stress is also sinusoidal and is given by:

$$\sigma(t) = \sigma_0 \sin(\omega t + \delta) \tag{10.9}$$

where δ is the phase angle or mechanical loss angle, and σ_0 is the stress amplitude. Furthermore, σ_0 at a given frequency is proportional to γ_0, if the strain is sufficiently small. This type of behavior is called linear viscoelasticity. While the linear viscoelastic properties could be described in terms of the amplitude ratio, σ_0/γ_0 and the phase angle, δ, it is customary to rewrite eqn (10.9) using a trigonometric identity as follows:

$$\sigma(t) = \gamma_0[G' \sin(\omega t) + G'' \cos(\omega t)] \tag{10.10}$$

where $G'(\omega)$ is the storage modulus and $G''(\omega)$ is the loss modulus, which are functions of frequency.

Capillary rheometers cannot be used to determine the viscoelastic properties, and rotational instruments must be used. While most rotational rheometers are designed solely for the measurement of viscosity, the Brabender Dynvimeter, an in-line sensor, and the Rheometrics On-Line Rheometer, are concentric-cylinder rheometers designed for the measurement of the storage and loss moduli.

10.2.5. Plasticity and the Yield Stress

Another type of rheological behavior that can occur in the case of concentrated suspensions is 'plasticity', which means that the material has a yield stress. Below the yield stress no flow occurs, while above the yield stress the material flows, exhibiting shear thinning or thickening and, often, structural time-dependency, e.g. thixotropy. This further complicates the measurement of rheological properties, particularly while a material is being processed.

10.3. VISCOSITY AS A TOOL FOR MATERIAL CHARACTERIZATION

In some applications, for example viscometers for heavy-fuel oil burners, it is the viscosity itself that one wishes to control, but more often the rheological property is used as a measure of some other material characteristic that is of primary interest. This material characteristic might be solvent concentration, extent of reaction or particle size. In such a case it is important to know the relationship between the rheological property measured, usually the viscosity, and the characteristic of interest. For example, if solvent concentration is of interest, one must know the relationship between viscosity and concentration, and this can be readily established in the laboratory. However, it is important to take into account other factors that can influence the viscosity of the solution, especially if those factors can vary during the normal operation of the process.

An important example of a competing factor is temperature. All rheological properties depend upon temperature; for example the viscosity of a liquid decreases rapidly as temperature is increased. An empirical equation often used to describe the variation of viscosity with temperature is the Arrhenius equation:

$$\frac{\eta(T)}{\eta(T_0)} = \exp\left[\frac{E_a}{R}\left(\frac{1}{T} - \frac{1}{T_0}\right)\right] \tag{10.11}$$

where $\eta(T_0)$ is the viscosity at a reference temperature T_0, and E_a is an activation energy for viscosity.

Consider, for example, the situation mentioned above, where viscosity is to be used as an indication of solvent concentration. The viscosity is a function of temperature as well as concentration:

$$\eta = \eta(C, T)$$

This dependency of viscosity upon concentration, C, and temperature has been determined in the laboratory. Since the temperature of the fluid in the process rheometer is subject to variation, this temperature must be measured and taken into account in order to make the correct quantitative interpretation of the output signal of the viscosity sensor in terms of solution concentration.

A common way of taking temperature effects into account in the interpretation of a viscosity signal is to carry out a temperature compensation. In other words, measure the fluid temperature at the point of measurement and use this information together with an empirical equation to calculate what the viscosity would be at some standard temperature, T_0. To illustrate this procedure, consider a case in which the variation of viscosity with temperature at constant concentration is given by the Arrhenius equation. Furthermore, let us say that over the range of concentrations likely to be encountered in the process, the variation of viscosity with concentration and temperature can be separated into a product of two functions:

$$\eta(T, C) = \eta(C, T_0) \exp\left[\frac{E_a}{R}\left(\frac{1}{T} - \frac{1}{T_0}\right)\right] \tag{10.12}$$

The signal from the process viscometer gives a value of the viscosity corresponding to the temperature in the viscometer, T_v. If this temperature is measured, then the viscosity at the same concentration but at a standard reference temperature, T_0, can be calculated as follows:

$$\eta(C, T_0) = \eta(C, T_v) \exp\left[\frac{E_a}{R}\left(\frac{1}{T_0} - \frac{1}{T_v}\right)\right] \tag{10.13}$$

This temperature-compensated viscosity can now be used as an indication of solution concentration for the purposes of process control. In other words, a control loop designed to maintain a constant value of concentration can accomplish this only by maintaining a constant value of $\eta(C, T_0)$.

Some commercial process rheometers have temperature compensation built in either by an electronic circuit or by means of software. These are usually based on an empirical curve-fitting procedure based on known viscosity values at two or three temperatures.

10.4. DESIGNING RHEOMETERS FOR PROCESS APPLICATIONS

10.4.1. General Requirements for a Process Rheometer

A rheometer that is to be used as a process sensor must address a number of important requirements in its design. There are practical issues dealing with the instrument's installation, use and maintenance. There are theoretical issues that directly affect the instrument's reliability. There are also important performance issues that affect an instrument's usefulness in real-time applications. A list of these requirements follows.

1. A process sensor must be robust, requiring little maintenance and must operate at the process pressures with negligible leakage. The plant environment is not at all like the laboratory, and delicate transducers will not last long.
2. It must not interfere with the process by influencing up- or downstream processing conditions. It must not alter the product or act as a hold-up spot where product can degrade or impurities can collect.
3. The sensor output signal should be directly related to a well-defined rheological property. For example, if a uniform shearing deformation can be generated and the corresponding shear stress monitored, the viscosity can be easily calculated. We shall see, however, that some of the simplest process rheometers involve poorly controlled or non-uniform deformations. In such a case, an accurate measurement of the viscosity cannot be made, and the output can only be used as a relative measure of material characteristics.

   This may or may not be a problem. As was explained earlier, viscosity is often used as an indicator of some other material characteristic. The effects of a fundamental inaccuracy in the viscosity measurement can be compensated for in the required viscosity–material-property correlation. As a matter of principle, however, it is always best to use a fundamentally sound measurement.
4. A sensor must produce a reliable (or repeatable) signal, reasonably free of noise. The plant environment is a noisy one in many senses. There is usually considerable mechanical noise, i.e. vibration, from a variety of sources, and this will introduce noise into the output of any mechanical transducer. There can also be considerable interference from electrical signal transmission, particularly in the case of low-level signals. This can arise from strong capacitance fields and

the starting and stopping of powerful machinery. These considerations suggest that amplification of a low-level signal should be carried out as closely as possible to the transducer generating the signal.

5. A sensor must react with a minimum of delay. Measurement delay is defined as the time that elapses from the moment that the process fluid passes the point in the process where the rheometer is installed to the moment an output signal is available that reflects the state of that fluid. For process control applications, it is crucial that the measurement delay be small. An empirical rule of thumb is that the measurement delay should be less than $1/10$ of the characteristic time of the process being controlled. An instantaneous measurement is the ideal, but, in practical terms the delay should be a matter of seconds rather than minutes. In the case of an on-line installation, the flow of the fluid in the sampling line and gear pump is often the major contributor to the delay. In many cases the sampling line is long, and the flow rate is low so that this transit time can be quite long. The transit time question is complicated somewhat by Taylor diffusion, which results from the fact that the velocity in a tube is not uniform across the diameter. The result of Taylor diffusion is that some of the material entering the transit line, i.e. that near the center, will travel much more rapidly to the rheometer than that near the wall. As a result of this, if there is a sudden change of viscosity of the fluid in the process, the rheometer will start to respond as soon as the fastest-travelling fluid reaches it, but this will be mixed with older fluid moving near the wall, and it will be some time before the signal from the rheometer reflects completely the change in viscosity of the process fluid. For this reason, it is an oversimplification to take the ratio of volume of the sampling line to the volumetric flow rate as a measure of the contribution of the transit time to the total measurement delay. Finally, once fluid actually enters the rheometer, the flushing of old fluid by new is not instantaneous, and this will further delay the response of the sensor. We will refer to this as the sample renewal problem. The basic challenge is to provide for a rapid rate of sample renewal without interfering with the process or compromising the reliability of the rheometer.

10.4.2. Special Requirements for Various Types of Fluid

10.4.2.1. Newtonian Fluids

It is relatively straightforward to determine the viscosity of a Newtonian fluid. However, problems can arise due to the level of viscosity of a material. If the viscosity is very low, as in dilute aqueous solutions, the stress level will be very low, leading to a poor signal-to-noise ratio. Another problem that can arise with low-viscosity materials is the occurrence of turbulence. The relationships between viscosity and the quantity actually measured, for example a pressure drop, a torque on a shaft, a piston transit time, etc., are all based on the assumption that the flow is laminar. If the flow in the viscometer is turbulent, these relationships are no longer valid, and the sensor is of no use for measuring the viscosity.

If the viscosity is very high, as in the case of molten plastics, turbulence will never be a problem, but the rheometer must be designed to accommodate very large stresses.

10.4.2.2. Non-Newtonian Fluids

Most commercial process rheometers are intended for use with Newtonian fluids. When the fluid being processed is non-Newtonian, special problems arise. Viscosity must be measured at a specific shear rate. Moreover, the viscosity at a single shear rate may not provide an adequate characterization of the state of the material; a series of tests at different strain rates might be needed. Materials that are often non-Newtonian include processed foods and polymer solutions and melts. In addition, the viscoelastic properties of these materials may be of interest in addition to or in place of the viscosity. Rotational rheometers are preferred for non-Newtonian fluids, as the output signal can be more easily interpreted in terms of a shear-rate-dependent viscosity or of the viscoelastic properties.

10.4.2.3. Suspensions

Suspensions of solids can be troublesome in several ways. Firstly, concentrated suspensions are likely to be non-Newtonian, exhibiting structural time-dependency and plasticity and making it difficult to obtain an unambiguous rheological characterization. In addition, it is important to prevent the build-up of solids on the surfaces of the rheometer. This is a particular concern if there is a small shearing gap in the rheometer where the collection of particles can impede the flow or erode the shearing surfaces.

It is sometimes useful to be able to monitor the viscosity of mineral slurries. The special requirements in such an application are that the instrument must not clog in service, should drain itself when flow stops, and must be designed so that it will not be damaged by particle abrasion. In addition, the particles may tend to settle out in the rheometer. Thomas et al.[2] have described an inverted U-tube capillary viscometer that meets these requirements. Reeves[3] has proposed a concentric-cylinder rheometer for use with mineral slurries in which particle settling is a problem. In Reeves' design, there is a constant overflow of slurry around the rim of the outer cylinder, which is drained from the bottom. An early approach to dealing with suspensions containing settling particles was simply to measure the torque required to rotate an impeller in the fluid.[4] However, care must be taken to avoid turbulent flow, and the correlation between torque and viscosity is only an approximate one.[5] A variation on this theme is the use of a helical screw impeller that rotates within a tube somewhat larger than the impeller,[5] and such a device has been used as an on-line viscometer for fermentation broths[6] and fluid food suspensions.[7]

10.4.2.4. Foods and Fermentation Broths

Processed foods and fermentation broths are usually suspensions of solids, and the problems discussed in the previous section are likely to arise.

Viscometers have been used to advantage for the continuous monitoring of fermentation processes. However, it is necessary that all exposed surfaces be sterilizable, and this eliminates most commercial process rheometers. Furthermore, fermentation broths are three-phase systems in which the presence of gas bubbles makes it very difficult to make a meaningful rheological measurement. While slit,[8] vibrational[9] and rotational[5,6] devices have been proposed for this type of material, none of these represents an ideal solution to the problem.

An important consideration with foods is that there be no dead zones in the rheometer where material can collect and degrade. Also, it must be possible to thoroughly clean all the surfaces to which the food is exposed. Steffe and Morgan[10] evaluated a commercial concentric-cylinder rheometer for determining the storage and loss moduli of an extruded rice dough. They found that the results were accurate but that there were numerous problems including: the high pressure required to pump the dough into the rheometer, dead zones where material could accumulate, and cleaning problems resulting from the difficulty of disassembly. Beer et al.[11] developed a concentric-cylinder rheometer expressly for use with food materials and evaluated its use with chocolate, which has a yield stress.

10.4.2.5. Polymeric Materials

It would be very desirable to be able to monitor some characteristics of the molecular-weight distribution of a resin as it emerges from a polymerization reactor. In addition, plastics resins are processed in the molten state to form them into final products, and it may be desirable to measure a rheological property of the melt as it is being processed. Several major problems arise in the on-line measurement of rheological properties of melts. Firstly, they are non-Newtonian, elastic and have a high viscosity, often in the range 10^3–10^5 Pa s (see Dealy and Wissbrun[12] for a complete treatment of the rheology of molten polymers). In addition, they are processed at elevated temperatures, usually in the range from 150 to 300°C. Finally, depending upon the application, the pressure may also be quite high, up to 10 000 psi (68 MPa).

Most of the currently available commercial process rheometers designed for use with melts are pressure-driven capillary or slit models that measure only an apparent viscosity (defined in Section 10.5.2) or a simulated 'melt index'. The melt index is a crude measure of the 'flow-ability' of a melt, as determined using a laboratory capillary viscometer of a very specific geometry.[13] There are standard test methods in many industrial countries that involve the use of such a 'melt indexer' or 'extrusion plastometer'. In the US, ASTM standard test method 1238 covers such a test. Because it is so widely used, those using process rheometers with molten polymers often wish to correlate the output of a process rheometer with the melt index. It is important to emphasize that the success of the apparent viscosity–melt index correlation is dependent upon how well the melt indexer geometry is reproduced by the process rheometer.

Capillary and slit rheometers that are designed to determine the variation of viscosity with shear rate[14] and elastic properties[15] have been described in the research literature, although commercial versions are not available.

Rotational rheometers for melts have been described by Heinz,[16] Kepes[17,18] and Starita and Macosko.[19] These have the advantage that the storage and loss moduli, which are linear viscoelastic properties, can be measured.

10.4.3. Requirements for High-Pressure Operation

High process pressures are another situation requiring special attention. Not all sensors can be sealed against high pressure. Process streams are usually at positive pressure, and in many applications the pressure is quite high. This poses two types of problem in rheometry: one is the necessity to provide dynamic seals for shafts used to transmit torque to a pump or rotating rheometer fixture, and the second is the need to determine the torque on one of the fixtures in the case of a rotational rheometer. Obviously, if the torque is measured on a rotating shaft, it will include a substantial contribution from seal friction, and this precludes its use for rheometric purpose. The solutions to this problem employed by commercially available instruments are outlined in Section 10.6.1.

The measurement of torque can be avoided altogether by using a shear-stress transducer that senses the local shear stress over a small area of one of the rheometer fixtures.[20] Alternatively, rheometers have been designed which use rotational motion to generate a high pressure in some region of the rheometer, which can then be easily measured. A relationship between this pressure and the viscosity of the fluid must then be established. This is the approach taken by Kraynik *et al.* in the 'helical screw rheometer',[21,22] which has been found useful with coal–solvent slurries, drilling muds and hydraulic fracturing fluids.[23]

10.5. CAPILLARY AND OTHER PRESSURE-FLOW RHEOMETERS

Capillary viscometers are the most popular of all rheometers, because of their simplicity in construction and use. The basic elements of a viscosity measurement consist of the measurement of the pressure drop for a given flow rate. Alternatively, one can fix the pressure drop and measure the flow rate. As is shown below, for flow in a capillary or slit this information is adequate for the determination of the viscosity of a Newtonian fluid. In the case of non-Newtonian fluids, the determinaton of a viscosity value is not straightforward, although capillary and slit rheometers may still provide useful information about such materials.

10.5.1. Primitive Pressure Drop Sensors
The pressure drop for flow past a flow constriction, for example a die or an orifice, is sometimes used as an indication of a fluid's viscosity. Pabedinskas *et al.*[14] used this approach to monitor the amount of degradation in a stream of molten polymer. There is a price to be paid for the simplicity of this approach. Firstly, the shear rate is highly non-uniform so it is not possible to derive a relationship between the pressure drop and the viscosity, particularly for non-Newtonian fluids. Such a correlation, if it exists, must be established experimentally. Secondly, the flow rate is governed by the needs of the process and not the rheological sensor. Moreover, this flow rate is generally not precisely known and varies with time owing to normal process upsets. Finally, the fluid temperature will

not be uniform across a large flow conduit, and this will further obscure the relationship between viscosity and the measured pressure drop.

10.5.2. Capillary Viscometers

10.5.2.1. *Basic Equations for Shear Stress, Shear Rate and Viscosity*

For fully developed flow in a tube, i.e. far from the entrance, the pressure gradient and the velocity profile do not change with distance, z, along the tube. By carrying out a force balance on a length, Δz, of tube, it can be shown that the shear stress at the wall, σ_W, is related to the pressures P_1 and P_2, at the upstream and downstream ends, respectively, of this length, and to the radius R of the tube as shown by eqn (10.14).

$$\sigma_W = \frac{(P_1 - P_2)r}{2\Delta z} \tag{10.14}$$

This arrangement is shown in Fig. 10.3. However, rather than measure pressure at two points in the fully developed flow region, the more common procedure is to measure only the driving pressure, P_d, in the reservoir feeding the tube. Then, if the pressure at the exit of the tube, i.e. at $z = L$, is atmospheric, and this is assumed to be small compared to P_d, an apparent wall shear stress can be calculated as shown in eqn (10.15).

$$\sigma_A = \frac{P_d R}{2L} \tag{10.15}$$

However, just downstream of the reservoir, the flow in the tube is not fully developed, and the pressure gradient is higher than that in the fully developed region. Thus, this apparent shear stress is not the wall shear

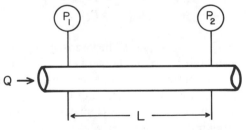

Fig. 10.3. Sketch of arrangement used to determine the pressure drop for fully developed flow in a tube. Volumetric flow rate is Q, tube radius is R, pressure at upstream measurement station is P_1, and pressure at downstream station is P_2. The distance between these stations is called Δz in eqn (10.14).

stress for fully developed flow, and in fact it is higher than the value given by eqn (10.14). If the excess pressure drop due to the entrance effect, ΔP_e, were known, the true wall shear stress could be calculated as follows:

$$\sigma_W = \frac{(P_d - \Delta P_e)R}{2L} \tag{10.16}$$

In process rheometers, this correction is generally not made, but if the ratio L/D is large, the correction is small.

There are other reasons for making L/D large, i.e. for making R small, which is why tube-flow rheometers are always 'capillary' rheometers. Firstly, for a given flow rate the driving pressure is larger, and this means a larger output signal for a given viscosity. Secondly, the temperature inhomogeneity resulting from viscous heating is reduced to a minimum. Finally, the probability of encountering turbulent flow is reduced. Turbulent flow occurs when the Reynolds number is larger than about 2100. Since the Reynolds number is equal to $(\rho \dot{\gamma} R^2/2\eta)$, making R smaller will make the Reynolds number smaller for a given shear rate and viscosity.

For a Newtonian fluid, we know that the velocity profile is parabolic, and from this the shear rate at the wall can be shown to be given by

$$\dot{\gamma}_W = \frac{4Q}{\pi R^3} \tag{10.17}$$

Thus, if Q is fixed and P_d is measured, the viscosity can be calculated from

$$\eta = \frac{\sigma_W}{\dot{\gamma}_W} = \frac{\pi R^4 P_d}{8LQ} \tag{10.18}$$

If the fluid is non-Newtonian, i.e. if the viscosity depends upon the shear rate, eqns (10.17) and (10.18) are no longer valid. There is a technique that can be used to determine the true wall shear rate in such a case, but it requires the differentiation of pressure data for a number of flow rates, and this is not practical for a process viscometer. In this case it is convenient to define an apparent wall shear rate as follows:

$$\dot{\gamma}_A \equiv \frac{4Q}{\pi R^3} \tag{10.19}$$

For the power-law fluid defined by eqn (10.7), the true wall shear rate is given by:

$$\dot{\gamma}_W = \left(\frac{3n+1}{4n}\right)\dot{\gamma}_A \tag{10.20}$$

The value of n can be significantly different from 1 for many materials of commercial importance, so the difference between $\dot{\gamma}_A$ and $\dot{\gamma}_W$ can be large.

Note that most capillary-type process viscometers give as an output signal an 'apparent viscosity' calculated as follows:

$$\eta_A \equiv \frac{\sigma_A}{\dot{\gamma}_A} = \frac{\pi P_d R^4}{8LQ} \tag{10.21}$$

A more detailed analysis of the capillary flow of non-Newtonian fluids is given in Refs 12 and 13.

10.5.2.2. Use of Capillary Viscometers as Process Sensors

If the fluid being processed is Newtonian and if L/D is large, the viscosity calculated using eqn (10.20) will be very close to the true viscosity. But for a non-Newtonian fluid the viscosity is dependent upon shear rate, and the true shear rate is not known. Thus, the output signal corresponds to the apparent viscosity at a given apparent shear rate. While such a signal may still be useful as an indication of fluid consistency, it is important to keep in mind that it is not the actual viscosity at some known shear rate. Another difficulty that can arise with non-Newtonian fluids is that the apparent viscosity at a single apparent shear rate may not provide sufficient information about the material. One way to obtain the apparent viscosity at more than one apparent shear rate is to use a pressure-driven device in which the fluid flows through a series of capillaries having different diameters.[24,25] Humphries and Parnaby[15] have proposed the use of a rheometer for molten plastics that consists of a three-section flow channel. A converging channel is followed by a straight section and then by a diverging section. It is claimed that this device can be used to determine a creep compliance and an apparent extensional viscosity.

It is important to note that in a capillary viscometer, the sampling rate is governed by Q, the flow rate in the viscometer, i.e. by the speed of the gear pump used to feed the viscometer. This flow rate is selected to optimize the performance of the viscometer and is usually rather low owing to the small diameter of the tube. But this implies a low sampling rate and thus a long measurement delay for the sensor. This difficulty can be circumvented by adding an additional gear pump to provide a bypass stream parallel to the capillary or slit,[26] although this adds to the complexity, size and cost of the device.

Thomas *et al.*[2] developed a capillary viscometer expressly for use with mineral slurries. It was desirable that the unit should not clog, that it should drain itself automatically on shut down, and that there be minimum effect of wear on calibration. Their solution to this problem was an inverted U-tube capillary viscometer. The bend in the tube affects the relationship between the wall shear stress and the pressure drop, and an empirical procedure was proposed for calculating the additional pressure drop due to the bend.

The Conameter Heavy Fuel Viscometer and the VAF Viscotherm are hybrid units with features of both in-line and on-line viscometers. These are designed primarily for the control of fuel oil viscosity in large combustion systems. The capillary and the gear pump are immersed in the oil flow inside a special flanged pipe fitting. The drive motor for the gear pump is mounted on the outside of the housing.

It was noted in an earlier section (10.4.2.5) of this chapter, that the most commonly used measure of melt consistency in the plastics industry is the melt index and that because of this, users of process rheometers in this industry often wish to interpret the signal from a process rheometer in terms of the melt index. We note that in a standard melt indexer, the driving pressure is generated by a weight-loaded piston, and the standard capillary has an L/D of about 4. This is a rather low value, and this means that a significant portion of the driving pressure is taken up by an entrance effect and that fully developed flow occurs over very little, if any, of the length of the capillary. For these reasons, the melt index depends upon rheological properties in addition to the viscosity, and the only way to get a reliable value for the melt index is to carry out the test exactly as specified in the standard test method. Thus, any formula used to convert the output signal of a process rheometer to a melt index must be verified for the melt of interest, by comparing the results of a standard melt indexer with those obtained from the process rheometer.

10.5.3. Slit Rheometers

A slit is a rectangular flow channel with a rather large ratio of width to height. For practical purposes the flow can be considered to be two-dimensional, which makes the equations needed to calculate rheological properties no more complicated than those for capillary flow. While slits are somewhat more complex to manufacture than capillaries, they have the advantage that pressure transducers can be easily mounted directly in the wall of the slit.

10.5.3.1. Basic Equations

The equations presented here are valid for a slit with a large aspect ratio, i.e. w/h, where w is the width and h the height. An aspect ratio of 10 is usually adequate for reasonable accuracy of the equations. For fully developed flow, i.e. away from the entrance, the magnitude of the wall shear stress is related to the pressure drop as follows:

$$\sigma_W = \frac{(P_1 - P_2)h}{2L} \tag{10.22}$$

As mentioned in the discussion of capillary viscometers, if the total driving pressure is measured rather than two pressures in the fully developed flow region, the pressure gradient corresponding to fully developed flow can only be determined if the entrance pressure drop is known. The situation is exactly analogous to that for a capillary rheometer.

For a Newtonian fluid, the magnitude of the shear rate at the wall is

$$\dot{\gamma}_W = \frac{6Q}{h^2 w} \tag{10.23}$$

Thus, for a Newtonian fluid, the viscosity is

$$\eta = \frac{\sigma_W}{\dot{\gamma}_W} = \frac{(P_1 - P_2)h^3 w}{12LQ} \tag{10.24}$$

10.5.3.2. Determination of the First Normal Stress Difference

Polymeric fluids are elastic, and the pressure drop measurements described above provide no information about fluid elasticity. A phenomenon exhibited by elastic fluids but not by Newtonian fluids, is the generation in a shear flow of 'normal stress differences'. The largest and most easily measured of these is the 'first normal stress difference', which is a measure of the degree of molecular orientation generated by the shear flow. If this quantity could be measured in a process rheometer, it would provide useful information about the polymeric material being processed. Because of this, there have been several attempts to develop such a rheometer. Three approaches have been taken. In two of these, Han's exit pressure method[27] and Lodge's hole pressure method,[28,29] a straight slit is employed, while in Geiger's method,[30,31] a curved slit is used. There are no commercial versions of any of these devices currently available.

Furthermore, there remains some uncertainty about the reliability of all three methods.

10.5.3.3. Slit Rheometers as Process Sensors

The difficulties that arise in using a slit rheometer for a non-Newtonian fluid are the same as those mentioned in the discussion of capillary viscometers. For example, eqns (10.23) and (10.24) are not valid, although an apparent viscosity can be calculated as $\sigma_W/\dot{\gamma}_A$. Furthermore, if the fluid is known to follow the power-law viscosity model and the power law index, n, is known, the true viscosity can be determined. To obtain information at more than one apparent shear rate, Fritz and Ultsch[32] have proposed the use of two slits having different gaps, h, arranged in series, with pairs of pressure transducers installed in both slits so that the apparent viscosity at two apparent shear rates can be determined. Pabedinskas et al.[33] have proposed the use of a 'wedge', i.e. a slit in which the value of h varies continuously along its length so that the apparent wall shear rate varies along the flow path. Then the wall pressure is measured at three locations, and if the fluid follows a power-law viscosity model, the power-law index can be determined. If such a device is used as an in-line rheometer, as proposed, we note that the entire process stream must pass through the slit, and this may not be desirable. Also, the flow rate is subject to normal process fluctuations and may not be precisely known.

10.6. ROTATIONAL PROCESS RHEOMETERS

10.6.1. Basic Principles

The usual mode of operation for a rotational rheometer is to set the rotational speed, Ω (radians per second), for the rotating fixture and measure the resulting torque, M, on either the stationary fixture or the rotating fixture itself. The most common 'geometries' for the fixtures are concentric cylinders, parallel disks and cone and plate. In all cases, for a Newtonian fluid, there is a linear relationship between these two quantities, as shown by eqn (10.25a):

$$M = \eta\Omega/K \tag{10.25a}$$

where K is an instrument constant that depends upon the detailed design of the fixtures. If this constant is known, either from a theoretical analysis or by calibrating the instrument with a Newtonian fluid of known viscosity, then the viscosity can be determined from eqn (10.25b).

$$\eta = KM/\Omega \qquad (10.25b)$$

For non-Newtonian fluids, the situation is generally more complex, as will be seen in the following detailed analyses.

The equations derived in this section are valid for flow in which all the streamlines are circles. This will be the case, for example, in a laboratory rheometer in which one places a sample in the instrument and leaves it there for the duration of a series of measurements. However, in a process instrument, one wishes to track the viscosity of a flowing steam of fluid, and this requires that the fluid in the rheometer be replaced either continuously or at frequent intervals. We have referred to this as the sample renewal problem. If there is a steady flow of fluid through the rheometer, to provide for continuous sample renewal, this may change the relationship between the measured torque and the viscosity.

10.6.2. Sample Renewal
When rotational rheometers are used as process sensors, sample renewal depends upon flow through the rheometer gap due to the pressure gradient in the process line or sampling circuit. A simple sample-renewal arrangement is shown in Fig. 10.4. For non-Newtonian fluids, it is desirable that the rheometer gap be quite small so that the shear rate is reasonably uniform. As a result, the rate of sample renewal is quite slow. If one tries to increase this by increasing the pressure flow, this will alter the velocity distribution in the rheometer and compromise its operation. This may make it necessary to stop the flow through the rheometer while a measurement is in progress.

In the Rheoprocessor invented by Kepes,[17,18] the stationary member is a bicone, and the sample stream enters the shearing gap by means of a hole in the center of one side of this bicone.

10.6.3. Sealing and Torque Measurement
The principal challenge in designing a rotational rheometer for use as a process sensor arises from the need to seal the device against the process pressure. Any type of dynamic seal will introduce significant friction, and this will be a major source of error if the torque is measured on the shaft that transmits the rotational motion from the drive system to the rotating cylinder. Various techniques have been used to overcome this problem. One of these involves the use of a 'torque tube' to support the stationary

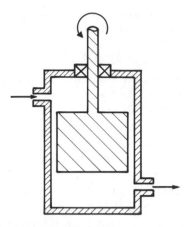

Fig. 10.4. Sketch of concentric-cylinder process rheometer. Sample renewal stream enters at upper left and exits at lower right. Torque must be measured either on the rotating shaft connected to the inner cylinder or on the fixed outer cylinder.

cylinder. This is a hollow tube that twists slightly in response to the torque transmitted to it by the fluid. This torsional deflection is tracked by means of a solid rod that passes through the center of the torque tube and out of the rheometer body. This approach has been used by H. K. Bruss[34] and others. Another alternative is the use of a magnetic coupling to transmit the torque into the rheometer body. Of course the sealing problem does not arise in the case of an immersion viscometer, and the torque can be measured directly on the drive shaft.

10.6.4. Viscoelastic Fluids

Rotational rheometers can, in principle, be used to determine the storage and loss moduli. Whereas the moving fixture rotates at constant speed to determine the viscosity, for measuring $G'(\omega)$ and $G''(\omega)$ it is driven in small-amplitude sinusoidal oscillation. Since the strain is small, the resulting torque is small, and this makes it difficult to obtain a favorable signal-to-noise ratio. Also, because of seal friction, it will be difficult to generate a true sinusoidal strain. In Chapter 7 it is shown how waveforms other than a sinusoid can be used to determine linear viscoelastic properties. A practical approach for process rheometry would be to generate a periodic shear strain that may be only roughly sinusoidal and then to monitor the true shape of the waveform actually generated and take this into account in the data analysis. It could even be advantageous to have a waveform rich in harmonics, as this could generate values of the storage and loss moduli at several frequencies simultaneously.

10.6.5. Concentric-Cylinder Rheometers

10.6.5.1. Basic Equations

We first present the equations for a case in which we do not take into account the shear stress acting on the ends of the inner cylinder. If the inner and outer cylinder radii are R_i and R_o respectively, and the length of the inner cylinder is L, the torque on the inner cylinder, M_i, is related to the shear stress at the wall of the inner cylinder, σ_i, for any type of fluid as follows:

$$M_i = 2\pi R_i^2 L \sigma_i \qquad (10.26)$$

For a Newtonian fluid, the shear rate at the wall of the inner cylinder, $\dot{\gamma}_i$, is related to the rotational speed as follows:

$$\dot{\gamma}_i = \frac{2\Omega}{[1 - (R_i/R_o)]} \qquad (10.27)$$

Making use of eqn (10.6), we can combine the above two equations to give the Margules equation for calculating the viscosity:

$$\eta = \frac{\sigma_i}{\dot{\gamma}_i} = \frac{M}{4\pi L \Omega} \left(\frac{1}{R_i^2} - \frac{1}{R_o^2} \right) \qquad (10.28)$$

The constant, K, of eqn (10.25) can be clearly identified.

If we now take into account the contributions to the torque by the shear stress acting on the ends of the cylinder, the torque will be larger, and eqn (10.28) will no longer be correct. However, the additional torque will generally still be proportional to the product of the speed and the viscosity so that eqn (10.25) can still be used if we determine K by use of a calibrating fluid of known viscosity.

It is important to note that all of the equations presented above for determining the viscosity by use of a concentric cylinder instrument are based on the assumption that the flow is laminar and that all the streamlines are circular. However, when the viscosity is low and the rotational speed is high, secondary flow or even turbulent flow can occur. When it is the outer cylinder that rotates, this is not likely to be a problem, but when the inner cylinder rotates, secondary flow will occur in a narrow-gap viscometer when

$$\left[\frac{\Omega R_i \rho (R_o - R_i)}{\eta}\right]\left(\frac{R_i - R_o}{R_i}\right)^{1/2} > 41$$

When secondary flow occurs, the shear rate pattern becomes more complex, and the viscosity can no longer be determined.

10.6.5.2. Non-Newtonian Fluids in Concentric Cylinder Fixtures

It is important to note that, even for a Newtonian fluid, the shear rate varies across the gap between the cylinders, and the shape of the curve of $\dot{\gamma}$ *versus* r depends upon the shape of the curve of viscosity *versus* shear rate. Thus, if $\eta(\dot{\gamma})$ is not known *a priori*, an explicit equation for the shear rate cannot be derived, and the viscosity cannot be determined. However, useful information can still be obtained. If, for example, the gap between the cylinders is small, i.e. if R_i/R_o is very close to one, the shear rate becomes nearly uniform and is given approximately by

$$\dot{\gamma} \approx \frac{\Omega R_i}{(R_o - R_i)} \tag{10.29}$$

For example, if (R_i/R_o) is $0\cdot99$, the variation in the shear rate across the gap is only 2%. Another advantage of using a small gap is that secondary flow or turbulence are less likely to occur. However, the use of a small gap may be undesirable in a process sensor because of the possibility of particulates sticking in the gap and clogging the rheometer. If the general form of the $\eta(\dot{\gamma})$ relationship is known, it may be possible to derive an explicit equation for calculating the viscosity when the gap is not small. For example, for a power-law fluid, the shear rate at the wall of the inner cylinder is given by:

$$\dot{\gamma}_i = \frac{2\Omega}{n[1 - (R_i/R_o)^{2/n}]} \tag{10.30}$$

However, this equation neglects end effects, and these are more difficult to deal with than in the case of a Newtonian fluid.

For non-Newtonian fluids, it is almost certain that the sample renewal flow will alter the relationship between the measured torque and the viscosity in a rather complex manner. Thus, when a concentric cylinder is used as a process sensor with a non-Newtonian fluid, one does not generally expect to obtain an accurate value of the viscosity of the fluid.

Instead, one obtains a number that is only approximately related to the viscosity but which may still serve as a useful measure of the consistency of the fluid.

Concentric-cylinder on-line sensors for molten polymers have been described in the literature in which either the outer[16,35] or the inner[31,36] cylinder rotates. Other rotational rheometers have been described above in the sections on food materials[11] and suspensions.[3]

10.6.5.3. Sample Renewal in Concentric Cylinder Viscometers

A further complication arises when there is a continuous flow of fluid through the shearing gap in order to provide for sample renewal. For a Newtonian fluid, if the sample renewal flow is primarily in the axial direction, the torque will still be proportional to the viscosity times the rotational speed. However, in a typical process rheometer, the flow pattern is generally more complex, and it will be necessary to carry out calibration experiments to determine reliably the relationship between the torque and the viscosity. Commercial instruments are usually precalibrated at the factory for a prescribed range of viscosities.

The use of axial flow to generate sample renewal, as shown in Fig. 10.4, is not practical in an *in-line* installation, as one would not wish the entire process stream to flow through the rheometer gap. This would be disastrous for the performance of the rheometer and would introduce a large pressure drop in the process. Thus, this technique can be used only in an *on-line* installation making use of a side stream. For in-line installations, the entire viscometer unit is immersed in the process stream, and slots in the outer cylinder permit some flow of fluid through the shearing gap. We note, however, that in this case there is no control of the sample renewal rate, and if the secondary flow due to flow through the slots becomes significant compared to the drag flow generated by the rotation of one of the cylinders, the response of the sensor will be compromised.

Recognizing the need to exercise direct control on sample renewal, Ferguson and Zhenmiao[37] developed a concentric-cylinder on-line rheometer in which two tiny piston pumps can be periodically activated to force fresh material into the rheometer gap. A torque tube is used to transmit the torque on the stationary cylinder (see Section 10.6.3).

Another solution to this problem involves the use of the rotating member of the rheometer to promote sample renewal by drag flow rather than pressure flow. This is the approach taken in the instrument sketched in Fig. 10.5. Since a shear-stress transducer[20] is used to monitor the shear stress, rather than inferring it from the total torque on a shift, only a

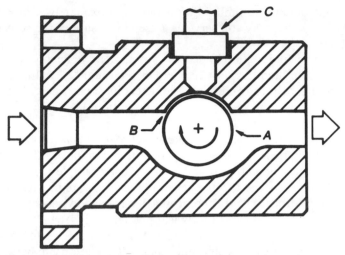

Fig. 10.5. An in-line concentric-cylinder rheometer. By using a shear stress transducer, C, rather than inferring the shear stress from a torque measurement, only a portion of the outer cylinder is required. The rotating cylinder, A, drags fluid into the shearing gap, B, to provide for sample renewal.

segment of the outer cylinder is required, and material can flow into this gap from the main flow.[38] However, even in this geometry the sample renewal process is not straightforward, because recirculation zones may be present in the rheometer.[39]

10.6.6. Parallel-Disk Rheometers
This flow geometry is illustrated in Fig. 10.6. The shear rate varies linearly with r and is given by

$$\dot{\gamma} = \Omega r/h \tag{10.31}$$

where h is the gap between the disks. For a Newtonian fluid, the viscosity is related to the torque as follows:

$$\eta = 2Mh/(\pi \Omega R^4) \tag{10.32}$$

If there is a cross flow through the gap for sample renewal, this renders the equation invalid. Neither is the equation valid for a non-Newtonian fluid because of the non-uniformity of the shear rate.

The BTG In-Line Viscosity Transmitter is a parallel-disk rheometer that is immersed in the process flow by means of a special flanged fitting.

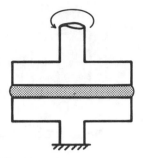

Fig. 10.6. Sketch of parallel-disk rheometer geometry. The fluid is sheared between a stationary, lower disk and a rotating, upper disk. From eqn (10.31) the shear rate varies with radial position.

Sample renewal is by means of radial slots in the moving disk. The BTG VISC-2000 Viscosity Transmitter is a concentric-cylinder unit in which an impeller on the drive shaft pushes fresh fluid into the shearing gap. A mechanism is provided to prevent large particles from entering the shearing gap.

Perry *et al.*[40] used a parallel-disk viscometer installed in a reaction-injection molding machine to track the viscosity of the reaction mixture.

10.6.7. Cone-and-Plate Rheometers

In the ideal geometry shown in Fig. 10.7 the apex of the cone just touches the plate without transmitting any torque to it. In practice the precise positioning of the fixtures to achieve this is difficult. Furthermore, if there is any particulate matter present, it will make contact with the fixtures near the apex, contributing to the torque and scoring the fixtures. For this reason, the central zone of the cone is usually machined flat, which introduces a small error into the equations presented here.

Cone-and-plate geometry is of special interest because the shear rate is approximately constant between the fixtures and is given by

$$\dot{\gamma} = \Omega/\Theta_0 \tag{10.33}$$

where Θ_0 is the cone angle, usually in the range of 5–8°. The shear stress is therefore also approximately uniform and is given by

$$\sigma = 3M/2\pi R^3 \tag{10.34}$$

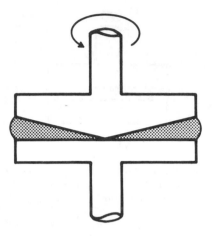

Fig. 10.7. Sketch showing cone-and-plate geometry. As shown by eqn (10.33), the shear rate is independent of radial position.

Thus, the viscosity is

$$\eta = 3M\Theta_0/2\pi\Omega R^3 \qquad (10.35)$$

Because of the uniformity of the shear rate, this equation is valid for non-Newtonian as well as Newtonian fluids. However, the presence of a cross flow resulting from continuous sample renewal will introduce an error that will increase with the speed of the cross flow.

10.7. HELICAL-FLOW RHEOMETERS

Helical-flow rheometers have been mentioned above in the discussions of measurements involving suspensions and high pressures. While a rotating member generates the shear, the principle of operation is somewhat different from that for the rotational instruments described in the previous section. In a helical-flow sensor a central shaft with a helical flight on it turns inside a cylindrical tube slightly larger in diameter than the flight. There are two modes of operation. In one, the torque required to rotate the shaft at a prescribed speed is measured. In this type of unit the flight is considerably larger than the shaft. It is designed for low-pressure operation, and the principal objective is to keep suspended matter from settling. Applications include fermentation broths[5,6] and processed foods.[7] The relationship between the measured quantities and the viscosity is considerably more complex than in the case of the rotational instruments described in the previous section, and there is no straightforward method for dealing with non-Newtonian materials. Kemblowski *et al.*[5] have developed approximate equations for calculating an 'average' shear rate and an 'average' shear stress.

In the second type of helical-flow rheometer, the shaft diameter is closer to that of the flight, and the cylinder is sealed at both ends. Instead of the torque on the shaft, it is the pressure difference between the ends of the cylinder that is measured. The principal objective is to permit measurements at high pressures. Since torque is not measured, seal friction is not a source of error. However, the device retains the ability to hold solids in suspension, and by periodically opening a valve to allow the device to act as a pump, sample renewal can be provided so that the device can serve for intermittent, on-line measurements. This instrument was first developed by Kraynik *et al.*[21,22] for use with coal–solvent slurries, but later found application with drilling muds and hydraulic fracturing fluids.[23] Equations for relating the measured pressure drop to the viscosity for Newtonian and power-law fluids have been presented by Kraynik *et al.*[22]

10.8. PISTON–CUP VISCOMETERS

When a piston located inside a larger cylinder moves axially, it subjects a fluid in the gap between these two elements to a shearing deformation. There are two basic types of fixture. In one, the outer cylinder is open at both ends, while in the other it is closed at one end, i.e. it forms a cup. Only the latter design is currently used as the basis of commercial process viscometers and, consequently, only it will be considered here.

This design, in which one end of the cylinder is closed, is illustrated in Fig. 10.8. During the measurement stroke, the piston moves toward the end of the cup, while during the sample renewal stroke it moves away from the end, and fresh fluid is drawn into the cup. During the measurement stroke, fluid displaced from the cup by the piston rises up near the wall of the cup, while near the piston, fluid is dragged along with it. This leads to a somewhat complex velocity profile, and the viscosity of a Newtonian fluid is given by

$$\eta = \frac{F[(R_o^2 - R_i^2)\ln(R_0/R_i) - (R_o^2 - R_i^2)]}{2\pi L V (R_o^2 + R_i^2)} \quad (10.36)$$

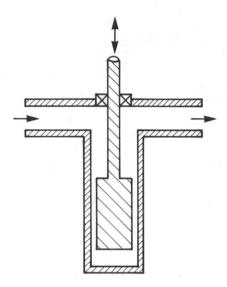

Fig. 10.8. Sketch of piston-cylinder process rheometer. The piston is alternately lifted, to allow for fresh fluid to fill the cup, and then lowered, to measure the viscosity. The time required for the piston to move a certain distance under the influence of a fixed driving force is proportional to the viscosity of a Newtonian fluid.

where R_i is the radius of the piston and R_o is the inner radius of the cylinder or cup. If the piston falls under its own weight, then the force is given by

$$F = \pi R_i^2 L (\rho_p - \rho_f) \tag{10.37}$$

where ρ_p and ρ_f are the densities of the piston and fluid, respectively. The time required for the piston to move a prescribed distance under the influence of a fixed force is inversely proportional to the velocity, and from eqn (10.36) this means that the transit time is directly proportional to the viscosity for a Newtonian fluid.

We note that as the piston falls, i.e. moves towards the closed end of the cup, it eventually displaces most of the fluid from it, and then draws fresh fluid into the cup when it is raised. In this way, the normal operation of the viscometer provides for good periodic sample renewal.

The Norcross Industrial Viscometers are of the falling-piston type. For low-viscosity fluids, the piston falls under its own weight for a measurement, the time of fall giving the velocity. It is then raised by an air cylinder preparatory for the next measurement. For high-viscosity fluids, the rod is also driven by an air cylinder during the measurement stroke to speed up the cycle.

In the small Cambridge Applied Systems sensors, the force required to move the piston is supplied by electromagnetic coils. The transit times for the motion of the piston in both directions are measured and averaged, and eqn (10.36) thus requires modification, although the viscosity is still proportional to the transit time.

10.9. VIBRATIONAL RHEOMETERS

A very attractive solution to the sample renewal problem is to use an in-line rheometer in which there is no shearing gap. An active fixture both generates the deformation and senses the fluid's response. This is exactly the method of operation of a vibrational viscometer. The response time is very short, and sample renewal is not a problem. In addition, there is no motor-driven element, and no dynamic seal is required, so operation at high pressure is not a problem. The sensor is compact and can be autoclaved if required. It is important to note that the operation of a vibrational viscometer always involves the acceleration of fluid. Because of this, the output signal is not simply proportional to the viscosity alone but to the product of the viscosity and the density. However, since the viscosity is usually much more sensitive to composition than the density, this is not normally a problem.

A vibrational viscometer is an example of a 'surface-loaded' rheometer, in contrast to a 'volume-loaded' or 'gap-loaded' instrument. This means that the deformation of the fluid is limited to a thin layer next to the viscometer fixture. It is essential that the surface of the vibrating member, which can be a rod, blade or sphere, be exposed to the process stream at all times. If there is any tendency for a layer of solids or degraded material to build up on the probe, the response of the instrument will be compromised. Furthermore, in the case of a multiphase system, the fluid very close to the probe may not have the same composition as the bulk fluid. Finally, in the case of non-Newtonian fluids, there is no direct relationship between the output signal and any well-defined rheological property.

While a 'vibrating-plate viscometer' was described by Woodward in 1951,[41] the first commercial unit was based on the theoretical work of Roth and Rich.[42] Their design involved longitudinal vibrations in a blade and formed the basis for the Bendix Ultraviscoson, which was the precursor of the Model 1800 Viscometer now offered by ABB Process Analytics. A magnetic pulse starts the vibrations in the blade, and when the amplitude has been damped by the fluid to a preset level, another pulse is automatically introduced. The frequency with which it is necessary to introduce pulses is a measure of the product of viscosity and density.

Another, simpler mode of operation for a vibrational viscometer is to provide a constant level of power to the coil generating the vibration and use the amplitude of the resulting vibration as a measure of the viscosity times the density. Because the amplitude must be within a certain range in order to be detected, a given configuration can only be used for fluids having viscosities within a fairly narrow range. The Dynatrol units made by Automation Products make use of a U-shaped vibrating element,[43] while the Sofraser viscometer made in France and sold in the US by Fisons Instruments has a rod that vibrates laterally.

The Nametre Viscoliner employs a circular probe, either a cylinder or a sphere, that is made to oscillate in a torsional mode.[44] The amplitude of the vibration is maintained constant by means of a feedback loop that provides sufficient power to the driver coil to accomplish this. This power is directly related to the product of viscosity times density. This makes the circuitry somewhat more complicated, but a major advantage is that a single unit can measure a very large range of viscosities. Also, the instrument is precalibrated and provides a direct, accurate reading of viscosity times density. Because of the feedback loop and associated amplifier, the system is sensitive to mechanical noise, which is always present in industrial systems. In the latest version of this instrument, a lock-in amplifier is incorporated into the circuit to maintain stable operation in spite of noise.

Because the shear rate is not uniform, there is no straightforward way of determining the true viscosity of a non-Newtonian fluid. However, vibrational viscometers may still be useful in applications involving such materials if an experimentally established correlation is available between the output signal and the material characteristic of central importance. Because the Nametre viscometer at least generates an axially symmetrical shear pattern, it is possible to extract information about the linear visco-elastic properties of a liquid by measuring the frequency shift between the driver and response waves, as well as the driver power.[45]

10.10. OTHER TYPES OF PROCESS RHEOMETER

We mention here two units that do not fit into any of the standard categories used above. In the Convimeter, made by Brabender, the moving element is a tapered cylinder mounted on a flexible tube. Inside this tube is an angled rod that is rotated by a motor. Thus, the tapered cylinder does not rotate on its axis but instead gyrates or wobbles to produce a sweeping motion. There is no rigorous theory of operation, and the relationship between torque and viscosity is determined empirically. The advantage of this design is that no dynamic seal is necessary, as the flexible tube provides the sealing. Thus, the unit can operate at 350 psi and at 550°C.

In the MBT moving-blade viscometer made by BTG, a flat blade moves from side to side inside a U-shaped housing that is open at the sides. The time required for the blade to move from one side to the other is measured and used to calculate the viscosity.

10.11. SELECTING A PROCESS RHEOMETER

The Appendix to this chapter consists of a table listing a number of manufacturers of process rheometers and their products. We tried to make the list as complete as possible, and any omission is unintended. Neither is inclusion intended to imply any endorsement or recommendation by the authors. In selecting an instrument for a particular application, we have seen that there are a number of factors that must be taken into account. The major factors and the related considerations for rheometer selection are summarized in this section.

Sensor performance specifications can be confusing and are frequently misused. The best procedure is to establish the level of performance required by a given application and to ask candidate manufacturers if they can meet these requirements.

10.11.1. The Nature of the Material Being Processed

10.11.1.1. Viscosity Level
Viscometer manufacturers often cite viscosity ranges as *zero* up to some maximum value. The lower limit of zero implies an infinite resolution, which is obviously impossible. The actual ratio between the maximum and minimum viscosities that can be measured with a particular configuration is usually in the range of 10–100, although this depends strongly upon the design of the unit. For example, it is highly unlikely that a unit with a maximum measurable viscosity of 100 Pa s will be able to cope

with a fluid having a viscosity of 20 centipoise (20 mPa s), even though the specifications may give the range as 0–100 Pa s.

Viscosity ranges may be specified in poise (P), centipoise (cP) or pascal-seconds (Pa s). The following conversions may be helpful:

$$1 \text{ cP} = 1 \text{ mPa s} = 0\cdot001 \text{ Pa s}$$

$$100 \text{ cP} = 1 \text{ P} = 0\cdot1 \text{ Pa s}$$

$$1000 \text{ cP} = 1 \text{ Pa s}$$

10.11.1.2. Presence of Particulate Matter

If solid particles are present in the fluid, they may damage the rheometer or interfere with its proper operation. This is especially important if there are small flow channels, for example in a capillary rheometer or in a small-gap rotational or piston–cup rheometer. Some manufacturers specify a maximum particle size that can be accommodated by each of their models.

10.11.1.3. Type of Rheological Behavior Exhibited

All the units listed in the appendix are suitable for use with Newtonian fluids. For non-Newtonian fluids, it is essential to know as much as possible about the actual rheological behavior of the fluid being processed before selecting a process rheometer. If the viscosity varies with shear rate, it is important to establish what rheological information is required to control the process. An apparent-viscosity signal generated by a very simple viscometer may be adequate for some applications, but in others the true viscosity at a specified shear rate may be required. Or the viscosity at two shear rates may be needed to make an estimate of the degree of shear thinning being exhibited by the fluid.

If viscoelastic properties are required, it is necessary to generate an oscillatory shear and interpret the results in terms of the dynamic viscosity or the storage and loss moduli. The Brabender Dynvimeter is designed for operation in precisely this way. The Nametre vibrational viscometer may also be able to provide some information about viscoelastic behavior.

10.11.1.4. Molten Plastics

Because of the special problems encountered in designing a rheometer for use with molten polymers, several manufacturers have specialized in this area. These include Rheometrics, Goettfert, Seiscor and Porpoise, all of whom make on-line capillary or slit rheometers. These are most often used to provide a signal that is related to the melt index of the material.

By mounting one of these units on the outlet of a small extruder, it is possible to sample a stream consisting of solid polymer in powder or pellet form. Porpoise makes a very compact unit for accomplishing this. The Brabender Dynvimeter and the Nametre vibrational viscometer are suitable for melts having relatively low viscosities.

10.11.1.5. Need for Sterile Processing

For use with food materials, cleanliness is essential, and stainless-steel construction will normally be specified. In addition, it is important that there be no zones in the flow field where material can collect and deteriorate. For pharmaceutical products and fermentation broths, sterility will be essential, and all components that come into contact with the process stream must be autoclavable.

If the rheometer is installed in a hazardous environment, an explosion-proof housing may be required. Many commercially available process rheometers are available in such a configuration.

10.11.2. Process Conditions

When a vendor specifies a 'maximum fluid temperature' it is usually higher than the 'maximum operating temperature'. Only those components immersed or in direct contact with the fluid will be subjected to the 'fluid temperature', while the components external to the flow will normally be at a lower temperature. Although heat will be conducted into the external components from those immersed in the flow, the former will also be cooled to some extent by the ambient air.

Note that the maximum operating pressure decreases with the operating temperature, because the mechanical performance and strength of metals and seals decrease with temperature. Therefore, the specification of a maximum pressure is not meaningful unless a temperature is also specified. If no temperature is given in the listing in the Appendix, the pressure specified should be assumed to be valid at ambient temperature.

There are two flow rates that may be of interest. In the case of an in-line installation, it is the flow rate in the process line in which the rheometer is mounted that is important. If this flow rate is too low, there may be inadequate sample renewal in the rheometer, while if it is too high, it may interfere with the operation of the rheometer. In the case of an on-line installation, it is the flow rate in the sampling side-stream that is important. If a gear pump is used to feed the sample stream, this flow rate will be governed by the speed of the pump. If a second pump is used to feed the fluid back into the process after it exits the rheometer, it should be operated at such a speed as to maintain the rheometer at some pressure that is well above the vapor pressure of the fluid but not so high as to exceed the mechanical specifications of the rheometer.

10.11.3. Process Time *versus* Signal Delay

For process-control applications, the characteristic time of the process is of central importance. It is essential that the measurement delay in the rheometer be short compared to this characteristic time. By measurement delay, we mean the time that elapses between the moment fluid passes the point where the rheometer is installed and the moment when a signal is available from the rheometer that is related to the state of that fluid. For fast processes, only a short delay can be tolerated. In-line units generally have a faster response than on-line units, and among the in-line units, the vibrational viscometers have the shortest delay—usually only a few seconds. In the case of on-line units, the transit time of the fluid in the sampling line makes a significant contribution to the measurement delay, so this line should be as short as possible. Some units produce an intermittent rather than continuous signal. In the case of piston–cup viscometers, there is a viscosity update only once each cycle. Indeed, this is the case whenever sample renewal is not continuous, for example in an on-line rotational rheometer where the feed pump is stopped while a measurement is being made.

REFERENCES

1. T. Hertlein and H.-G. Fritz, *Kunststoffe*, 1988, **78**(7), 606.
2. A. D. Thomas, N. T. Cowper and P. B. Venton, *Proc. Xth Int. Congr. of Rheology*, Vol. 2, University of Sydney, Sydney, Australia, 1988, p. 326.
3. T. J. Reeves, *Trans. Inst. Min. Met. Sec. C*, 1985, **94**, C201.
4. A. B. Metzner and R. E. Otto, *A.I.C.E. J.*, 1957, **3**, 3.
5. Z. Kemblowski, J. Sek and P. Budzynski, *Rheol. Acta*, 1988, **27**, 82.
6. Z. Kemblowski, P. Budzynski and P. Owczarz, *Rheol. Acta*, 1990, **29**, 599.
7. M. S. Tamura, J. M. Henderson, R. L. Powell and C. F. Shoemaker, *J. Food Sci.*, 1989, **54**, 483.
8. O. Neuhaus, G. Langer and U. Werner, *Chem.-Ing.-Tech.*, 1982, **54**, 1188.
9. D. Picque and G. Corrieu, *Biotechnol. and Bioeng.*, 1988, **31**, 19.
10. T. F. Steffe and R. G. Morgan, *J. Food Proc. Eng.*, 1987, **10**, 21.
11. T. Beer, H.-W. Suess and H. D. Tscheuschner, *Wiss. Z. Tech. Univ. Dresden*, 1988, **37**, 9.
12. J. M. Dealy and K. F. Wissbrun, *Melt Rheology and its Role in Plastics Processing*, Van Nostrand Reinhold, New York, 1990.

13. J. M. Dealy, *Rheometers for Molten Plastics*, Van Nostrand Reinhold, New York, 1982.
14. A. Pabedinskas, W. R. Cluett and S. T. Balke, *Polym. Eng. Sci.*, 1989, **29**, 993.
15. C. A. M. Humphries and J. Parnaby, *Proc. Inst. Mech. Engrs*, 1986, **200**, 325.
16. W. Heinz, *Proc. IXth International Congress on Rheology*, 1984, **4**, 85.
17. A. Kepes, *Rheology* (Proc. 8th Int. Congr. Rheol.), G. Astarita, M. Matruci & L. Nicolais (Eds), Vol. 2, Plenum Publishing Co., New York, 1980, p. 185.
18. A. Kepes, US Patent No. 4 334 424 (1982).
19. J. M. Starita and C. W. Macosko, *Society of Plastics Engineers* (*ANTEC*) *Technical Papers*, 1983, **29**, 522.
20. J. M. Dealy, US Patent No. 4 463 928 (1984).
21. A. M. Kraynik, J. H. Aubert and R. W. Chapman, in *Proc. IXth Int. Congr. on Rheology*, Vol. 4, UNAM, Mexico City, 1980, p. 77.
22. A. M. Kraynik, J. H. Aubert, R. N. Chapman and D. C. Guyre, *Society of Plastics Engineers* (*ANTEC*) *Technical Papers*, 1984, **30**, 405.
23. D. L. Lord and D. Shackelford, *J. Can. Petrol. Technol.*, 1990, **29**(3), 47.
24. J. L. Scheve, W. H. Abraham and E. B. Lancaster, *Ind. Eng. Chem. Fundam.*, 1974, **13**, 150.
25. G. B. Froishtcter, A. M. Manoilo, K. K. Triliskii and R. M. Manevich, 'A device for monitoring rheological properties of lubricating greases in a flow', in *Nov. Reol. Polim., Mater. Vses Simp. Reol. 11th Meeting,* Vol. 2, G. V. Vinogradov and L. Ivanova (Eds), 1980, p. 117.
26. A. Goettfert, *Society of Plastics Engineers* (*ANTEC*) *Technical Papers*, 1991, **37**, p. 2299. See also *Kunststoffe*, **81**, (1), 1991, 44.
27. C. D. Han, Slit Rheometry, in *Rheological Measurement*, A. A. Collyer and D. W. Clegg (Eds), Elsevier Applied Science, London, 1988, Ch. 2.
28. A. S. Lodge, Normal Stress Differences from Hole Pressure Measurements, in *Rheological Measurement*, A. A. Collyer and D. W. Clegg (Eds), Elsevier Applied Science, London, 1988, Ch. 11.
29. H. X. Vo, Doctoral Thesis, Chemical Engineering, University of Wisconsin, 1988.
30. H.-G. Fritz, *Kunststoffe*, 1985, **75**, 785.
31. G. Menges, W. Michaeli, C. Schwenzer and L. Czybooora, *Plastverarbeiter*, 1989, **40** (4), 207.
32. H.-G. Fritz and S. Ultsch, *Kunststoffe*, 1989, **79**(9), 785.
33. A. Pabedinskas, W. R. Cluett and S. T. Balke, *Polym. Eng. Sci.*, 1991, **31**, 365.
34. H. K. Bruss, US Patent No. 2 518 378 (1950).
35. R. D. Orwell, *Advances in Polym. Technol.* 1983, **3**(1), 23.
36. G. M. Khachatryan, K. D. K'yakov, A. A. Strel'tsov and K. N. Sosulin, *Khimicheskie Volokna*, no. 3, May–June 1983, p. 48.
37. J. Ferguson and X. Zhenmiao, Paper presented at 1988 Annual Meeting of British Society of Rheology, Surrey, submitted to *Rheol. Acta*.
38. T. O. Broadhead, B. I. Nelson and J. M. Dealy, *Internat. Polym. Proc.* (in press) 1992.
39. B. I. Nelson, T. O. Broadhead, J. M. Dealy and W. I. Patterson, *Internat. Polym. Proc.* (in press) 1993.

40. S. J. Perry, J. M. Castro, and C. W. Macosko, *J. Rheology*, 1985, **29**, 19.
41. J. G. Woodward, *J. Colloid Sci.*, 1951, **6**, 481.
42. W. Roth and S. R. Rich, *J. Appl. Phys.*, 1953, **24**, 940.
43. W. B. Banks, US Patent No. 3 292 422 (1966).
44. J. V. Fitzgerald, F. J. Matusik & D. W. Nelson, US Patent No. 3 382 706 (1968).
45. J. V. Fitzgerald, F. J. Matusik, D. W. Nelson and J. L. Schrag, US Patent No. 4 754 640 (1988).

APPENDIX—MANUFACTURERS OF COMMERCIAL INSTRUMENTS

ABB Process Analytics
PO Box 831,
Lewisburg, WV 24901,
USA.
(304) 647-1501

● Model 1800 Viscometer

   Vibrational viscometer, measures $\rho\eta$ product
   Maximum product: 5000 cP × g/cc
   Maximum operating pressure: 750 psi
   Maximum operating temperature: 300°C
   Response time: 3–5 s for 95% change
   Typical applications: fuel oil, varnishes, adhesives, inks

Automation Products Inc.,
3030 Max Roy Street,
Houston, TX 77008,
USA.
(713) 869-1485

● Dynatrol Viscosity Control

   U-tube vibrational viscometer; measures $\rho\eta$ product
   For given unit, max./min. value = 100
   Overall range for all units: $1-10^5$ cP
   Pressure rating: 3000 psig at 100°F
   Temperature rating: 300°F
   Explosion-proof rating

C. W. Brabender Instruments, Inc.,
50 Wesley Street,
South Hackensack, NJ 07606,
USA.
(201) 343-8425

- Convimeter
  Conical probe makes circular motion but does not rotate, so no
    seals or rotating parts are in contact with the fluid
  Viscosity ranges: many, from 0·15 to 250 Pa s full scale
  Maximum pressure: 150 psi (special version 750 psi)
  Maximum fluid temperature: 300°C

Brabender Messtechnik KG,
PO Box 35 01 62
D-4100 Duisberg 1,
Germany.
(0203) 770593

- Dynvimeter
  Concentric-cylinder immersion rheometer
  Uses oscillatory shear to determine viscoelastic properties
  For use with melts, pastes, etc.
  Dynamic viscosity range: $0·2–10^4$ Pa s
  Dynamic modulus range: 2·5–20 000 Pa
  Maximum fluid temperature: 280°C (special version, 350°C)
  Maximum pressure: 50 bar

- Immersion Viscometer

  Concentric-cylinder unit; inner cylinder rotates
  Overall viscosity range for all models: 5 mPa s to 60 Pa s

Brookfield Engineering Laboratories, Inc.,
240 Cushing Street,
Stoughton, MA 02072,
USA.
(617) 344-4310

- Viscosel Automatic Viscosity Control

  Rotating cylinder, for vented, immersion installation
  Plastic sample chamber with overflow available
  Typical applications: inks, coatings, adhesives
  Maximum viscosity for given configuration: 35 cP up to 2000 cP
  Maximum flow rate in sample chamber: 1·0 gpm

● TT100 In-Line Viscometer

 Concentric-cylinder viscometer with torque tube
 Sample renewal by axial flow with bypass
 Viscosity ranges:
  Maximum: 50–500 Pa s, depending upon model
  Minimum: 10–30 cP, depending upon model
  Maximum pressure: 500 psig
  Temperature range: −40 to 500°F

BTG Lausanne S.A.,
PO Box 236,
CH-1001 Lausanne,
Switzerland.
021/20 11 61

● MBT Moving-Blade Viscosity Transmitters
 Viscosity range:

  MBT-150: 30–200 000 cP
  MBT-155: 10–200 000 cP

 Maximum particle size: 1·0 mm
 Operating pressure: 10 bar (150 psi)
 Maximum fluid temperature:

  MBT-150: 85°C
  MBT-155: 150°C

 Maximum flow velocity: 2 m/s
 Explosion-proof installation available

● VISC-2000 Rotating-Cylinder Viscosity transmitter

 Immersion unit; impeller provides for sample renewal
 For use with highly contaminated and abrasive media; screening
  mechanism protects gap
 Viscosity ranges: 5–120 cP or 15–350 cP (Newtonian fluid)
 Maximum operating pressure: 10 bar (145 psi)
 Maximum fluid temperature: 150°C
 Electric and pneumatic models available

- VISC-21 Rotating-Disk Viscosity Transmitter

    Typical application: preheating heavy fuel oil
    Maximum fluid temperature 180°C
    Viscosity range: 0·008–0·7 Pa s
    Operating pressure: 80 bar (1200 psi)
    Electric and pneumatic models available

Cambridge Applied Systems, Inc.,
57 Smith Place,
Cambridge, MA 02138,
USA.
(617) 576-7700

- Electromagnetic Viscometer

    Piston–cup viscometer with electromagnetic drive
    In-line installation in small line by use of a 'T' fitting
    Immersion installation directly into a tank or large line
    Overall viscosity range for all pistons: 0·2 mPa s to 20 Pa s
    Max./min. viscosity for given piston: 20
    Maximum particle size: 25–360 microns, depending upon range
    Maximum operating pressure. 1000 psi at 70°C
    Maximum operating temperature: 190°C
    Typical applications: adhesives, coatings, paints, fuel oils

Conameter Corporation,
9 Democrat Way,
Gibbsboro, NJ 08026,
USA.
(609) 783-7675

- Heavy-Fuel Viscometer
    Immersed capillary viscometer
    Maximum viscosity: 200 or 300 Saybolt universal seconds
    　　　　　　　　　　　 200 Redwood 1 seconds
    Maximum flow rate: 100 gpm
    Maximum pressure: 500 psi

Fisons Instruments,
53 Century Road,
Paramus, NJ 07652-1482,
USA.
(201) 265-7865

● Sofraser MIVI Process Viscometer

    Vibrating-rod viscometer, measures $\rho\eta$ product
    Typical applications: paints, inks, fuel oil, adhesives
    Maximum operating temperature: 300°C
    Maximum operating pressure: 150 bar
    Lowest range: 0·5–100 mPa s
    Highest range: 5–1000 Pa s
    Output signal non-linear with viscosity; calibration curve needed.

Fluid Data Ltd.,
20 Bourne Industrial Park,
Bourne Road,
Crayford, Kent,
England, UK, DA1 4BZ

● Isoviscous Temperature Process Analyser
    Capillary viscometer
    Determines temperature at which fluid has a given viscosity
    For use with bitumen and fuel oil

Göttfert Werkstoff-Prüfmaschinen GmbH,
Postfach 1261,
D-6967 Buchen/Odenwald,
Germany.
(06281) 4080

● Bypass Rheograph and RTR On-Line Rheometer
    Capillary viscometers for molten plastics

Mettler-Toledo AG,
CH-8606,
Greifensee,
Switzerland.
(01) 944 22 11

● Covimat Concentric-Cylinder Viscometer

    Immersion and on-line models
    Magnetic drive
    Many viscosity ranges available
    Range covered by all models: 10^{-3}–5×10^3 Pa s
    High temperature and pressure versions available

Nametre Company,
101 Forrest Street,
Metuchen, NJ 08840,
USA.
908-494-2422

- Viscoliner Process Control Viscometer

    Vibrational viscometer, measures product $\rho\eta$
    Rod or sphere in torsional vibration
    Temperature range: -40–$500°C$
    Viscosity–density product range:
    (single unit covers entire range shown)
    Spherical sensor: 0.1 to $200\,000$ cP-g/cm^3
    Rod sensor: 100 to $2\,000\,000$ cp-g/cm^3

Norcross Corporation,
255 Newtonville Avenue,
Newton, MA 02158,
USA.
(617) 969-3260

- Industrial Viscometers; piston–cup type

    Models for immersion and in-line installation
    Air cylinder raises piston; for measurement stroke, both gravity
    and air-driven models available
    Max./min. viscosity for given unit: 20
    Viscosity ranges (full scale): 20–1000 Pa s
    High-pressure model max. pressure: 600 psi
    High temperature model max. temperature: 500°C

Porpoise Viscometers Ltd,
Peel House, Peel Road,
Skelmersdale, Lancs,
England, UK, WN8 9PT
(0695) 50002

- P3 On-Line Viscometer

    Capillary viscometer for molten polymers and other high-viscosity
    materials
    Gives simulated melt-index signal: 0.1–200 g/min

- P5 On-Line Process Control Viscometer

    Samples polymeric powder or granules
    Simulates melt indexer; range: 0·1–200 g/10 min

Rheometrics, Inc.,
One Possumtown Road,
Piscataway, NJ 08854,
USA.
(908) 560-8550

- Melt Flow Monitor (MFM)

    On-line slit rheometer with two gear pumps for use with molten
      plastics
    Can be mounted directly on top of extruder or die
    Overall viscosity range, all models: $100–2 \times 10^5$ Pa s
    Melt-index range, all models: 0·10–130 g/10 min
    Temperature range: 150–350°C

- Rheometrics On-Line Rheometer (ROR)

    Concentric-cylinder rheometer; outer cylinder rotates
    Can be used to measure storage and loss moduli of molten plastics

Seiscor Technologies, Inc.,
PO Box 470580,
Tulsa, OK 74147-0580,
USA.
(918) 252-1578

- CMR-II By-Pass Rheometer

    On-line capillary rheometer
    Simulates melt indexer
    Viscosity range: $1–10^5$ Pa s
    Melt index range: 0·2–1000 g/10 min
    Temperature range: 100–350°C
    Pressure range: 35–3000 psi
    Shear rate range: $0·1–7500 \text{ s}^{-1}$
    Polymer flow rate: 275–550 g/h

- RSR-100 Return-Stream Rheometer

    Returns melt to process
    Specifications similar to CMR-II

● Flow Characterization Rheometer

Polymer flow divided between two capillaries. If capillaries have different diameters, can measure apparent viscosity at two apparent shear rates simultaneously
Other specifications similar to CMR-II

● REX-1000 Automatic Sampler and Extruder System

For sampling flake or pellet
Consists of an extruder and an on-line capillary rheometer
Sampling is by ball valve, vacuum system and transfer line

Toray Engineering Co. Ltd.,
Instrument Marketing Department,
4-18, Nakanoshima 3-chome,
Kita-ku,
Osaka 530,
Japan.

● On-Line Capillary Viscometers, models VCM, VCH, and VCP

In some units, capillary is bent into form of U for compactness
Model VCM-5 designed for tank mounting
Model VCP-6 can be mounted on extruder
Model VCP-7 accepts solid polymer as feed
Full-scale viscosity: 1–1000 P
High-viscosity units: 150 Pa s to 2 kPa s
Maximum operating pressure: 20 kg/cm^2 or 200 kg/cm^2
Maximum operating temperature: 40°C or 250°C

VAF Instruments BV,
PO Box 40,
3300 AA Dordrecht,
The Netherlands.

● Viscotherm Fuel Control Unit

Immersed capillary viscometer; primary application is for fuel-oil viscosity control on board ships and in power plants
Magnetic coupling for gear-pump drive eliminates fire hazard
Dampening section in flow channel reduces problems due to pressure fluctuations

INDEX